Physical Acoustics and Metrology of Fluids

The Adam Hilger Series on Measurement Science and Technology

Series Editors: **M N Afsar, A E Bailey, R B D Knight** and **R C Ritter**

Other books in the series

SQUIDs, the Josephson Effects and Superconducting Electronics

J C Gallop

Uncertainty, Calibration and Probability: The Statistics of Scientific and Industrial Measurement

C F Dietrich

Physical Acoustics and Metrology of Fluids

J P M Trusler

Department of Chemical Engineering
Imperial College London

CRC Press
Taylor & Francis Group
Boca Raton London New York

CRC Press is an imprint of the
Taylor & Francis Group, an **informa** business

CRC Press
Taylor & Francis Group
6000 Broken Sound Parkway NW, Suite 300
Boca Raton, FL 33487-2742

First issued in paperback 2019

© 1991 by Taylor & Francis Group, LLC
CRC Press is an imprint of Taylor & Francis Group, an Informa business

No claim to original U.S. Government works

ISBN-13: 978-0-367-40303-4

Library of Congress Cataloging-in-Publication Data

Catalog record is available from the Library of Congress

Visit the Taylor & Francis Web site at
http://www.taylorandfrancis.com

and the CRC Press Web site at
http://www.crcpress.com

Contents

Preface

This book is about the theory and practice of experiments designed to measure the speed and absorption of sound in fluids. The results of such experiments are interesting because, from them, one can learn something about the macroscopic and/or microscopic properties of the fluid; this fact, which forms the central theme behind these pages, has motivated most experimenters in the field. Indeed, in recent years, very accurate measurements of the speed of sound in both gases and liquids have been made for the purposes of determining fundamental physical quantities and thermodynamic properties of fluids. Measurements of sound absorption have also contributed significantly to the microscopic and macroscopic physics of fluids. The science, both theoretical and experimental, behind such experiments is the primary focus of this book. Nevertheless, much of the theory is also relevant as background information for those interested in applications of acoustic measurements outside of experimental science.†

One of the great attractions of acoustic measurements is that they can often be performed with very high precision leading to sensitive measurements of some physical quantities that influence the speed and absorption of sound as functions both of frequency and of the variables that define the thermodynamic state of the fluid. If we are to exploit fully the very high precision with which quantities such as the resonance frequency of a fluid-filled cavity, or the time of flight of a sound pulse between two transducers, can be measured then we must deploy appropriate theory capable of describing the detailed performance of the chosen experimental method. A thorough understanding of the theory can guide the experimentalist towards better and more accurate methods and much of what follows will be directed towards this very important consideration.

It is appropriate to set the metrological problem in context by examining first what applications exist for acoustic experiments in pure and applied science; that is the objective of the first chapter. The basic acoustic theory

† Examples include the use of acoustic techniques to study the temperature of ocean water (e.g. as an indicator of global warming) or to measure temperature distributions inside industrial furnaces, and the use of sound propagation as the basis of a composition detector in gas chromatography.

is then developed in Chapter 2, mostly from first principles, with the primary objective of relating the speed and absorption of sound to the physical properties of the fluid medium in which it propagates. In Chapter 3, the same theory is applied to compute the steady-state sound field inside a closed cavity driven by a source of sound. This situation forms the basis of experimental methods in which the resonance frequencies of a fluid-filled cavity are used to infer the speed of sound in that fluid and the damping of the resonances is used to obtain information about the absorption of sound. Chapter 4 is concerned with processes, other than those of viscous damping and thermal conduction (the ever-present mechanisms discussed in Chapter 2), that influence the absorption of sound in some fluids. These include the important cases of vibrational, rotational and chemical relaxation. Chapter 5 is the interface between the theoretical and practical aspects of the subject; it is devoted to the generation and detection of sound. Those who have worked in the field will know that transducers are about the most problematic pieces of apparatus involved. Furthermore, since much of the interest in acoustic results refers to conditions of temperature and pressure remote from the ambient, experimenters often face severe constraints upon the materials from which transducers can be fabricated. Again, theory is used to lay down the guidelines along which devices ought to be designed but much practical information is also provided. Finally, in the last two chapters, we come to the main experimental methods. Steady-state and transient methods are described separately and both are illustrated by examples of practical instruments taken from the literature. These examples have been chosen to exemplify the best that can be achieved even though less sophisticated apparatus can often give satisfactory results.

The subject matter for this book draws together many areas of physical science including fluid mechanics, thermodynamics, kinetic theory, wave motion and the mathematical techniques of each of these fields as well as the host of practical skills necessary to design and execute good experiments. The reader unfamiliar with some or all of these subjects will be unlikely to master the subject solely from this book; he or she will need to refer to texts on some of the background subjects. However, I hope that there is enough information here to make clear that particular synthesis of ideas that forms the basis of accurate measurements.

Acknowledgments

I am most grateful for all the help and encouragement that I have received from family, friends and colleagues along the route leading to this book. I mention in particular Dr M B Ewing and Dr M R Moldover who have both contributed greatly to my knowledge and enjoyment of the subject.

During the preparation of the typescript, I have been assisted by the comments of Drs M Al-Laith, M B Ewing, S M Richardson, G Saville and V Vesovic and by Professor J B Mehl—all of whom I thank. Copies of original illustrations were kindly provided by Drs A R Colclough (figure 6.5), P J Kortbeek (figure 7.5) and B E Gammon (figure 6.7). Numerical data on the propagation of sound in rarefied monatomic gases, previously published only in graphical form, was provided for me from the notebooks of the late Dr M Greenspan by Dr D G Eitzen of the US National Institute of Standards and Technology; those data are reproduced in figure 4.1. Finally, I would like to thank my family for their unfailing encouragement and support during the writing of these pages.

J P Martin Trusler
London, April 1991

1

Fluid Properties from the Speed and Absorption of Sound

1.1 Introduction

In this opening chapter we examine the information that sound can provide for us about the medium through which it passes. That issue provides the rationale for acoustic measurements in experimental investigations of fluid properties. It will be addressed under three main headings. First, the speed of sound depends primarily on thermodynamic properties of the fluid and, under favourable conditions, can provide useful and very accurate information about thermodynamic quantities. Second, the absorption of acoustic energy throughout the bulk of a homogeneous fluid depends primarily upon kinetic effects relating to the distribution of energy on a molecular scale and, also under favourable conditions, can provide a measure of one or more time constants characterizing molecular processes. Finally, the reflection of sound waves from a boundary between the fluid and a solid surface is accompanied by phase changes and energy losses that can sometimes be used to infer the shear viscosity and thermal conductivity of the fluid with useful accuracy. We shall examine each of these in turn. In the course of this chapter, a few of the results of later chapters need to be anticipated so that the relations between acoustic quantities and those in which we are ultimately interested may be quantified.

1.2 Thermodynamic Properties from the Speed of Sound

The thermodynamic aspects of the subject, which will not be developed greatly in the chapters that follow, are examined here in rather more detail

1

than the other topics touched on in this introduction.[1] The emphasis is placed on determination, from the speed of sound, of the thermodynamic equation of state in the form of a relation between density, temperature and pressure. When combined with knowledge of the heat capacity as a function of temperature, which is also available from the speed of sound, such an equation is sufficient for the calculation of all the thermodynamic properties of the fluid. We shall see that the speed of sound alone is formally insufficient to determine the equation of state uniquely (additional constants of integration are required). However, it does provide extremely useful information which in combination with other experimental quantities, or with results obtained from molecular theory, can be used to construct the full equation of state.

We begin with a general relation, to be proved rigorously later, that provides a bridge between the speed of sound and thermodynamic properties of the fluid. This is then specialized to the cases of gases, liquids and near-critical fluids.

1.2.1   General expressions

The fundamental relation between the speed of sound u and thermodynamic properties of the fluid is

$$u^2 = (\partial p/\partial \rho)_S \qquad (1.2.1)$$

where p is the pressure, ρ is the mass density and S denotes entropy. As we shall see later, this expression is exact in the limits of small amplitude and low frequency. The former limit can always be achieved while the latter is usually, but not always, approached in practice. For our present purposes, we shall assume either that equation (1.2.1) is obeyed exactly or that appropriate corrections are known.

Since it is often inconvenient to retain the isentropic partial derivative, equation (1.2.1) is usually resolved into isothermal and either isobaric or isochoric terms.[2] When temperature and pressure are taken as the independent variables, standard thermodynamic manipulations can be used to show that

$$u^2 = [(\partial\rho/\partial p)_T - (T/\rho^2 c_p)(\partial\rho/\partial T)_p^2]^{-1} \qquad (1.2.2)$$

where T is the thermodynamic temperature and c_p is the isobaric specific heat capacity. Alternatively, temperature and density may be taken as the independent variables, in which case u^2 may be expressed as

$$u^2 = [(\partial p/\partial \rho)_T + (T/\rho^2 c_V)(\partial p/\partial T)_\rho^2] \qquad (1.2.3)$$

where c_V is the isochoric specific heat capacity.

Expressions for the heat capacities will also be required before u can be related to an equation of state in the form $\rho = \rho(T, p)$ or $p = p(T, \rho)$. For example, we may write

$$c_p(T, p) = c_p^{\ominus}(T) + \int_{p^{\ominus}}^{p} (\partial c_p / \partial p)_T \, dp \qquad (1.2.4)$$

where $c_p^{\ominus}$ is the value of c_p at a reference pressure $p^{\ominus}$. The partial derivative $(\partial c_p / \partial p)_T$ which appears here is given most conveniently by the thermodynamic relation

$$(\partial c_p / \partial p)_T = -T(\partial^2 \rho^{-1} / \partial T^2)_p. \qquad (1.2.5)$$

For gases and supercritical fluids, the reference pressure is usually taken as zero so that $c_p^{\ominus}$ becomes identical with the specific heat capacity of the perfect gas which we denote by c_p^{pg}. For liquids, the reference pressure is usually taken as either 0.1 MPa or the vapour pressure at the given temperature. In a similar manner, c_V can be expressed as the sum of a reference value $c_V^{\ominus}(T)$ and an isothermal integration of its derivative with respect to density; the derivative required in this case is given by

$$(\partial c_V / \partial \rho)_T = -(T/\rho^2)(\partial^2 p / \partial T^2)_\rho. \qquad (1.2.6)$$

Often it will be convenient to use molar heat capacity, $C_{p,m} = Mc_p$ or $C_{V,m} = Mc_V$, in place of specific heat capacity, and amount-of-substance density $\rho_n = \rho / M$ in place of mass density, where M is the molar mass.

The speed of sound can also be expressed in terms of a so-called canonical equation of state such as the molar Gibbs energy as a function of T and p, $G_m = G_m(T, p)$, or the molar Helmholtz energy as a function of T and ρ_n, $A_m = A_m(T, \rho_n)$. These thermodynamic surfaces contain complete information about all of the thermodynamic properties of the fluid so that auxiliary equations for the heat capacity are not required. The Gibbs energy is appropriate when the independent variables are temperature and pressure and this surface has the property that each of the three quantities, G_m, T and p, are equal in coexisting phases of a pure fluid. The Helmholtz energy is appropriate when temperature and density (or volume) are the independent variables and is often the most convenient choice when constructing an equation of state. For a Gibbs surface, one can show that

$$u^2 = \left(\frac{-(1/M)(\partial G_m / \partial p)_T^2 (\partial^2 G_m / \partial T^2)_p}{(\partial^2 G_m / \partial T^2)_p (\partial^2 G_m / \partial p^2)_T - [\partial(\partial G_m / \partial p)_T / \partial T]_p^2} \right) \qquad (1.2.7)$$

while in terms of a Helmholtz surface the speed of sound is given by

$$u^2 = \left(\frac{(\partial^2 A_m / \partial \rho_n^2)_T (\partial^2 A_m / \partial T^2)_\rho - [\partial(\partial A_m / \partial \rho_n)_T / \partial T]_\rho^2}{(M/\rho_n^2)(\partial^2 A_m / \partial T^2)_\rho} \right). \qquad (1.2.8)$$

1.2.2 The gas phase

It is convenient to develop the general expressions for u separately for gas and liquid phases, leaving the supercritical fluid to be taken up by either description as appropriate. The simplest case is that of the perfect gas for which the mass density is just Mp/RT (R is the gas constant). When substituted in equation (1.2.2) this leads to a simple expression for the speed of sound in a perfect gas:

$$u^2 = RT\gamma^{pg}/M \qquad (1.2.9)$$

in which $\gamma^{pg} = C_p^{pg}/(C_p^{pg} - R)$ is the zero-pressure (perfect-gas) limit of C_p/C_V. This predicts that the speed of sound in a gas is independent of pressure.

Since all gases become perfect at sufficiently low pressures, equation (1.2.9) must be correct in the limit $p \to 0$ but, at non-zero pressures, the intermolecular forces present in a real gas may be expected to cause some smooth departure from the predictions of the simple model. Anticipating the results of later discussion, we therefore write u^2 for a real gas as an expansion about the perfect-gas limit in powers of the pressure:

$$u^2 = A_0(T) + A_1(T)p + A_1(T)p^2 + \cdots. \qquad (1.2.10)$$

Here, the coefficients $A_0, A_1, A_2 \cdots$ are functions of temperature but not of pressure. We already know that A_0 is given by

$$A_0 = RT\gamma^{pg}/M \qquad (1.2.11)$$

and we can hope to obtain a precise measure of this quantity by extrapolation of sound speeds measured at a number of pressures along an isotherm to the limit of zero pressure. Although simple, this observation is the basis of several important applications of acoustic measurements. Once A_0 is known, we can obtain one of the four quantities on the right-hand side of equation (1.2.11) if we know the appropriate combination of the other three. For example, absolute measurements of the speed of sound have been used to determine the universal gas constant with exceptionally high accuracy [1, 2]. In the most recent measurement by this method [2], the gas chosen was a sample of argon, for which γ^{pg}/M was known with high accuracy, and the measurements were performed at the temperature of the triple point of water where T is known by definition. The overall fractional standard deviation in R was estimated at just 1.7 PPM (parts per million).

Another important application is in the field of primary thermometry where measurements of the speed of sound are used to infer the thermodynamic temperature [3–5]. From equations (1.2.10) and (1.2.11), we see that the zero-pressure limit of u^2 at some constant temperature T yields the value of that temperature directly if $R(\gamma^{pg}/M)$ is known. Monatomic gases have found favour because, except at extremely high temperatures where electronic degrees of freedom become active, there are no internal molecular contributions to the heat capacity of the gas and γ^{pg} takes the value 5/3 independent of temperature. A further advantage of monatomic gases is that they can be obtained with a high degree of chemical purity. For example, acoustic measurements using helium gas have made a major contribution to the establishment of practical temperature scales below 20 K [3, 4]. If the experiment is also performed at the temperature T_0 of the triple point of water using the same monatomic gas then the ratio $A_0(T)/A_0(T_0)$ will give T/T_0 without knowledge of $R(\gamma^{\mathrm{pg}}/M)$ being required. This is useful first because it eliminates the uncertainty in R and second because, although the molar masses of individual isotopes are usually known with very high accuracy, the isotopic compositions of commercially available gases are not always known unambiguously.

More generally the temperature is determined by measurement on a practical scale and the zero-pressure limit of the speed of sound is used to obtain the molar perfect-gas heat capacities $C_{p,\mathrm{m}}^{\mathrm{pg}}$ and $C_{V,\mathrm{m}}^{\mathrm{pg}}$ through the relation

$$C_{p,\mathrm{m}}^{\mathrm{pg}}/R = \gamma^{\mathrm{pg}}/(\gamma^{\mathrm{pg}} - 1). \qquad (1.2.12)$$

Acoustic measurements offer one of the most precise methods available for the determination of these quantities. Typically, $C_{p,\mathrm{m}}^{\mathrm{pg}}$ may be obtained for a nominally pure gas with a precision of 0.01 per cent or better [6, 7]. For many substances, the absolute accuracy is then controlled by the presence of impurities which influence A_0 through both γ^{pg} and M.

For a binary gaseous mixture of components A and B, the quantity γ^{pg}/M is given by

$$\gamma^{\mathrm{pg}}/M = [(R + C_{V,\mathrm{A}}^{\mathrm{pg}})/C_{V,\mathrm{A}}^{\mathrm{pg}} M_\mathrm{A}] \left(\frac{1 + a_1 x}{1 + a_2 x + a_3 x^2} \right)$$

$$\xrightarrow[x \to 0]{} [(R + C_{V,\mathrm{A}}^{\mathrm{pg}})/C_{V,\mathrm{A}}^{\mathrm{pg}} M_\mathrm{A}] [1 + (a_1 - a_2)x] \qquad (1.2.13)$$

where x is the mole fraction of B, and the coefficients a_i are given by

$$a_1 = (C_{V,\mathrm{A}}^{\mathrm{pg}} - C_{V,\mathrm{B}}^{\mathrm{pg}})/(R + C_{V,\mathrm{A}}^{\mathrm{pg}})$$
$$a_2 = (M_\mathrm{B}/M_\mathrm{A}) + (C_{V,\mathrm{B}}^{\mathrm{pg}}/C_{V,\mathrm{A}}^{\mathrm{pg}}) - 2 \qquad (1.2.14)$$
$$a_3 = [1 - (C_{V,\mathrm{B}}^{\mathrm{pg}}/C_{V,\mathrm{A}}^{\mathrm{pg}})][1 - (M_\mathrm{B}/M_\mathrm{A})].$$

Here $C_{V,\mathrm{A}}^{\mathrm{pg}}$ and $C_{V,\mathrm{B}}^{\mathrm{pg}}$ denote the values of $C_{V,\mathrm{m}}^{\mathrm{pg}}$ for the two components. Since γ^{pg}/M is usually a single-valued function of x, one application of equation (1.2.13) is in the determination of composition in binary mixtures. For example, sound speed measurements have been used to monitor composition in experiments designed to study the ordinary and thermal diffusion coefficients in binary mixtures [8]. The second form of equation (1.2.13) can also be used to calculate the effect of impurities on the speed of sound in a dilute gas. The heat capacities of some hydrocarbon gases have been obtained by measuring the change in u^2 accompanying the introduction of a small but known amount of the substance into an inert 'carrier' gas [9]. This has allowed $C_{p,m}^{pg}$ to be measured for substances that are too involatile for study as pure components in the gas phase.

There is also much to be learnt from the speed of sound about the thermodynamic properties of real fluids away from the perfect-gas limit. In particular, a casual inspection of equation (1.2.1) suggests that the density should be obtainable as a function of temperature and pressure from an integration of $u^{-2}(T, p)$. This is not a trivial observation; accurate measurement of $\rho(T, p)$ remains one of the long-standing problems in experimental thermodynamics and accurate prediction of that quantity is still an important and largely unsolved problem in statistical thermodynamics. Imperfect knowledge of the density of compressed gases is one of the limiting factors in gas-phase flow metering and it is therefore of commercial as well as scientific consequence. Despite many investigations, there are still numerous important systems, especially mixtures, for which our knowledge is either incomplete or insufficiently accurate.

Since the speed of sound can be measured with very high precision and is apparently much less susceptible to the sources of systematic error that plague conventional (p, ρ, T) measurements, the possibility of obtaining the density from $u^{-2}(T, p)$ is certainly an attractive one. The difficulty, as a second glance at equation (1.2.1) will reveal, is that u^{-2} gives directly $(\partial\rho/\partial p)$ not along as isothermal path as we would wish but on an isentropic path. Thus, before proceeding with a more detailed analysis, it is instructive to examine the general form of the isentropes. Figure 1.1 shows on a temperature–pressure diagram isentropic lines calculated from the van der Waals equation of state for a diatomic fluid with $\gamma^{pg} = 1.4$; the results for other fluids, and for more realistic equations, will not be qualitatively different. We see that the isentropic paths follow radically different directions in the gas and liquid phases. In the liquid, they rise almost vertically from the saturation curve while in the very dilute gas the isentropes are almost horizontal and approach the temperature axis as tangents. This difference has important implications for the extraction of the relevant portions of the (p, ρ, T) surface from speed of sound measurements. A typical rectangular space within which $u^{-2}(T, p)$ results for a gas might be obtained is illustrated in the figure. A numerical integration of these results will require boundary or initial values along a line that cuts across the isentropes and, from the general direction of those lines, we see that initial values are required along the lowest isotherm. An important but disappointing feature is that we cannot exploit our knowledge of the equation of state in the perfect-gas limit for this purpose.

To proceed, we can employ equation (1.2.2) or (1.2.3) so that only isothermal and either isobaric or isochoric partial derivatives are involved;

equation (1.2.2) is the appropriate choice when temperature and pressure are the independent variables. For a gas, it is convenient to eliminate ρ in favour of the more slowly varying compression factor $Z = Mp/\rho RT$ in terms of which

$$u^{-2} = (M/RTZ^2)\{[Z - p(\partial Z/\partial p)_T] - (R/C_{p,\mathrm{m}})[Z + T(\partial Z/\partial T)_p]^2\}$$

$$(1.2.15)$$

and

$$(\partial C_{p,\mathrm{m}}/\partial p)_T = -(R/p)[2T(\partial Z/\partial T)_p + T^2(\partial^2 Z/\partial T^2)_p]. \quad (1.2.16)$$

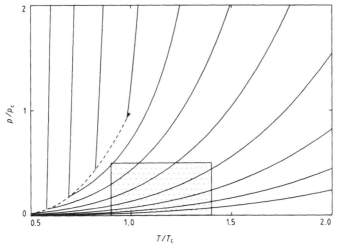

Figure 1.1 Isentropic lines calculated from the equation of state of van der Waals for a diatomic fluid with $\gamma^{\mathrm{pg}} = 1.4$, plotted as functions of reduced temperature T/T_c and reduced pressure p/p_c: ———, isentropes; – – – –, vapour pressure curve.

This shows that to obtain Z, and hence the density, inside a rectangular region in which u^{-2} is known we must solve a non-linear second-order differential equation subject to specified initial conditions. Integration of $u^{-2}(T, p)$ to obtain $Z(T, p)$ is a straightforward numerical procedure provided that initial values of Z and $(\partial Z/\partial T)_p$ are specified along the isotherm at the lowest temperature T_0. A simple Euler method is satisfactory, although others may be more efficient. From the initial conditions, $(\partial Z/\partial p)_T$ can be evaluated at T_0 and the acoustic data used to obtain $C_{p,\mathrm{m}}(T_0, p)$ from equation (1.2.15); then, differentiating $C_{p,\mathrm{m}}$ with respect to pressure, the second derivative $(\partial^2 Z/\partial T^2)_p$ can be evaluated as a function of p along the isotherm using (1.2.16). This provides sufficient information to estimate Z and $(\partial Z/\partial T)_p$ for a new isotherm at $T = T_0 + \delta T$ by means of Taylor series expansions about $T = T_0$ correct to the second derivative; further iterations of the whole procedure allow the surface to be obtained across the range covered by $u^{-2}(T, p)$. The method is stable in the sense that the effects of small errors in the initial conditions decay with increasing temperature along isobars.

In practice, numerical integration methods have not been widely employed with gases but alternative approaches to the problem have been pursued. The usual method is to postulate some explicit functional form for the equation of state that contains a set of adjustable parameters, the values of which are optimized using speed of sound results either alone or in combination with other thermodynamic properties of the fluid. Standard non-linear fitting methods can be used for this purpose. Often, in constructing an equation of state, T and ρ_n are adopted as the independent variables in place of T and p even though the latter are usually the experimental quantities. Equation (1.2.3) is now the appropriate starting point from which we obtain

$$u^2 = (RT/M)\{[Z + \rho_n(\partial Z/\partial\rho_n)_T] + (R/C_{V,\mathrm{m}})[Z + T(\partial Z/\partial T)_\rho]^2\} \quad (1.2.17)$$

and

$$(\partial C_{V,\mathrm{m}}/\partial\rho_n)_T = -(R/\rho_n)[2T(\partial Z/\partial T)_\rho + T^2(\partial^2 Z/\partial T^2)_\rho]. \quad (1.2.18)$$

Calculation of $u^2(T, p)$ will then involve first solving the equation of state for the density corresponding to the specified temperatures and pressures.

Generally, the accuracy of an equation of state developed using speed of sound data alone depends crucially on the choice of the functional form. Empirical forms, such as cubic equations, and very flexible multiparameter forms cannot be used with any accuracy unless other experimental information is incorporated to ensure that the solution for Z or ρ_n obtained satisfies appropriate boundary conditions [10]. However, if the equation has a firm theoretical basis then there may be sufficient constraints to allow the parameters to be determined using the speed of sound alone. For example, a theoretically based perturbation equation of state, containing a small number of parameters optimized using precise speed of sound results only, was shown to give excellent results for helium over a wide range of temperature and pressure [11].

The virial equation of state

$$p/\rho_n RT = 1 + B\rho_n + C\rho_n^2 + D\rho_n^3 + \cdots \quad (1.2.19)$$

is of special importance because it may be derived from molecular theory without significant approximation. The virial coefficients $B, C, D, \cdots$, which depend only on temperature in a gas of fixed composition, are related by the theory to the functions that describe the intermolecular potential energy of molecular clusters containing two, three, four, $\cdots$, molecules. In particular, the second virial coefficient B is related to the potential energy function U for a pair of molecules and when that depends only on the distance r between the molecules, B is given according to classical mechanics by the equation

$$B = 2\pi L \int_0^\infty [1 - \exp(-U/kT)] r^2 \, dr \qquad (1.2.20)$$

in which L is Avogadro's constant and k is Boltzmann's constant.[3]

When equation (1.2.19) is substituted in (1.2.17) and (1.2.18), and coefficients of powers of ρ_n are collected, the following series expansion for u^2 is obtained:

$$u^2 = (RT\gamma^{pg}/M)(1 + \beta_a\rho_n + \gamma_a\rho_n^2 + \delta_a\rho_n^3 + \cdots). \qquad (1.2.21)$$

The coefficients $\beta_a, \gamma_a, \delta_a, \ldots$, which we call acoustic virial coefficients, are also functions only of temperature for a gas of fixed composition. They are related to the corresponding coefficients and all the lower-order terms in equation (1.2.19), by second-order differential equations; for example, the second acoustic virial coefficient is given by

$$\beta_a = 2B + 2(\gamma^{pg} - 1)T(dB/dT) + [(\gamma^{pg} - 1)^2/\gamma^{pg}] T^2 (d^2B/dT^2) \qquad (1.2.22)$$

and the third is given by [12]

$$\gamma_a = [(\gamma^{pg} - 1)/\gamma^{pg}]$$
$$\times [B + (2\gamma^{pg} - 1)T(dB/dT) + (\gamma^{pg} - 1)T^2(d^2B/dT^2)]^2 + (1/\gamma^{pg})$$
$$\times \{(1 + 2\gamma^{pg})C + [(\gamma^{pg})^2 - 1] T(dC/dT) + \tfrac{1}{2}(\gamma^{pg} - 1)^2 T^2 (d^2C/dT^2)\}. \qquad (1.2.23)$$

B and β_a are illustrated for argon as functions of temperature in figure 1.2.

An analogous expansion of the compression factor in powers of p can also be taken as the starting point from which equation (1.2.10) may be obtained. The coefficients of equations (1.2.10) and (1.2.21) can be inter-related; for example,

$$(M/\gamma^{pg})A_1 = \beta_a \qquad (1.2.24)$$

$$(M/\gamma^{pg})A_2 = (\gamma_a - B\beta_a)/RT. \qquad (1.2.25)$$

The acoustic virial coefficients can also be related directly to intermolecular potential energy functions using integral equations. Thus, for example, if the pair potential is known then both β_a and B can be obtained from equations (1.2.20) and (1.2.22). Conversely, β_a could be used to infer information about $U(r)$. Indeed, the direct inversion procedure by which $U(r)$ can be obtained from $B(T)$ has been adapted to obtain the same function from $\beta_a(T)$ [13]. The advantage of the acoustic quantity is that it can be measured with greater accuracy, especially at temperatures below the critical where adsorption complicates direct measurements of B.

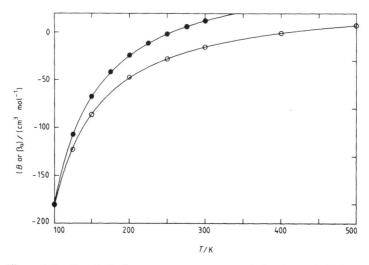

Figure 1.2 B and β_a for argon: $\circ$, recommended values of B from [28]; $\bullet$, experimental values of β_a from [29]. The full curves show values of B and β_a calculated from the expression $B/(\text{cm}^3\,\text{mol}^{-1}) = 178.93 - 142.59 \exp(93.1\,\text{K}/T)$.

Even though application of the virial equation linearizes the relation between u^2 and Z it does not in general eliminate the requirement for initial values. Each significant acoustic virial coefficient will require an initial value of the corresponding (p, ρ_n, T) virial coefficient and its first (or second) temperature derivative; that information is equivalent to knowledge of Z and $(\partial Z/\partial T)_\rho$ along an isotherm. If such information is available then numerical integration starting from the lowest temperature of interest is a straightforward and stable procedure [7]. Unfortunately it is at low temperatures that direct measurements of virial coefficients become most difficult. However, the great merit of the virial equation is that we have clear theoretical guidance as to the functional form of the first few virial coefficients. The inversion procedure mentioned above provides one possible route to solving equation (1.2.22) without recourse to any measurements of B itself. Since results over a rather wide temperature range are required for inversion, a similar but less rigorous approach using model potential energy functions has been employed in practice. Quite crude models with parameters determined using measurements of β_a have been found to give excellent results [6, 14]. Not only do they fit the data within experimental uncertainties but several different models which each provide a good representation of β_a have been found to give almost identical values of B even when extrapolated to temperatures substantially above the experimental range. Empirical polynomial expressions for B with coefficients determined using β_a also seem to give accurate results. In principle, the same approach can be applied to the higher virial coefficients. The third virial coefficient appears to be somewhat more sensitive to the shape of the pair potential and is also affected by intermolecular forces unique to three-molecule interactions. This means that empirical pair potentials that give both β_a and B with good accuracy cannot be relied upon to do the same for C and γ_a (or the corresponding coefficients of p^2 in the pressure-explicit expansions). Nevertheless, it seems likely that reasonably good solutions for C can be obtained from γ_a or A_2 in many cases using empirical expressions of the correct general form. It is always wise to test the hypothesis that a particular solution is close to the correct one by employing several different but plausible functional forms and then comparing the results.

1.2.3 The liquid phase

In liquids, measurements of the speed of sound offer one of the most accurate routes to obtaining the pressure dependence of the density. Although the absolute accuracy of measurements in dense fluids is usually slightly less than that which is possible in the gas, results approaching 0.01 per cent accuracy are now becoming available [15]. The accuracy of the sound speed measurements can be maintained even at pressures up to several hundred megapascal. Since direct measurements of density and isobaric heat capacity are relatively easy at atmospheric pressure but difficult at elevated pressures, a combination of direct and acoustic methods offers a well balanced approach of determining thermodynamic properties of liquids. In this combination, the density and heat capacity of the liquid at elevated pressures are obtained by numerical integration of $u^{-2}(T, p)$ subject to initial values obtained at a low pressure by direct measurement [15, 16].

The initial values required are ρ and c_p along a line that cuts across the isentropes; typically the isobar at atmospheric pressure is chosen below the normal boiling temperature but values along the saturation curve will do just as well. Assuming the former, we then know also $(\partial\rho/\partial T)_p$ and $(\partial^2\rho/\partial T^2)_p$ on the initial isobar allowing us to obtain $(\partial\rho/\partial p)_T$ from equation (1.2.2) and $(\partial c_p/\partial p)_T$ from (1.2.5). Then values of ρ and c_p can be obtained along an isobar at a higher pressure using Taylor series expansions truncated after the first derivatives. This simple Euler-type integration was first applied to results for mercury [16]; since then improved numerical techniques have been described [17]. The accuracy of such methods is good. The results give all the interesting thermodynamic properties of the liquid including density, isobaric expansivity, isobaric and isochoric heat capacities, isothermal and isentropic compressibilities, and enthalpy and free-energy increments.

The overall accuracy of the density is generally controlled by the accuracy of the initial values. Absolute measurement of ρ to around 0.01 per cent is not too difficult at atmospheric pressure but, since the method also requires the first and second derivatives with respect to temperature, the thermal expansion must be measured carefully. For typical organic liquids the expansivity, $(-1/\rho)(\partial\rho/\partial T)_p$, is around $1 \times 10^{-3}\,\mathrm{K}^{-1}$ and can be measured to better than 1 per cent; similar accuracy may be obtained for c_p at atmospheric pressure. Together these two quantities determine the second term in equation (1.2.2) allowing the important first term, $(\partial\rho/\partial p)_T$, to be obtained with an accuracy of at least 1 per cent. Significantly higher accuracy may be possible. The variation of c_p with pressure is usually quite small and the initial values of $(\partial^2\rho/\partial T^2)_p$ need not be very accurate. The overall accuracy of the derived density for typical liquids is about 0.1 per cent or better at pressures up to order 100 MPa [17]. This compares favourably with the results of other methods.

The speed of sound in a liquid is, like that in a gas, sensitive to temperature and could be used as the basis for a thermometer. For example, in liquid water at 300 K the value of $(1/u)(\partial u/\partial T)_p$ is about $1.8 \times 10^{-3}\,\mathrm{K}^{-1}$ which is similar to that in a dilute gas at the same temperature [18]. Since in the present case we have no *a priori* means of relating u to the thermodynamic temperature, a liquid cannot be used as the working fluid of a primary thermometer. Nevertheless, sensitive secondary thermometers could be constructed once u is known as a function of T under the appropriate conditions. The temperature of large volumes of liquid (ocean water, for example[4]) could be measured in this way using the propagation times of sound pulses along paths of known length in the liquid. The sound speed can also be related to composition in binary liquid mixtures or solutions when appropriate calibration measurements have been performed.

1.2.4   Critical behaviour

Near to the critical point of a pure fluid many of the thermodynamic and transport properties approach either zero or infinity. Such critical phenomena are now rather well understood and may be described in terms of more or less sophisticated scaling equations [19]. In particular, the isochoric heat capacity and the isothermal compressibility both diverge according to power laws that depend upon the path along which the critical point is approached. For example, approaching the critical temperature T_c from above along the critical isochore $\rho = \rho_c$ we have, for sufficiently small values of $T - T_c$,

$$c_V \sim |T - T_c|^{-\alpha} \tag{1.2.26}$$

$$(\partial\rho/\partial p)_T \sim |T - T_c|^{-\gamma} \tag{1.2.27}$$

where α and γ are positive constants known as critical exponents.

According to the three-dimensional lattice gas model of critical behaviour, which is in general agreement with experimental observations for real fluids, these exponents take the values $\alpha \approx 0.11$ and $\gamma \approx 1.24$ [20]. Similar power-law expressions apply below T_c for each of the coexisting phases (i.e. along the saturation curve) but possibly with slightly different values for the critical exponents. Along the critical isotherm, the divergence of c_V and of $(\partial\rho/\partial p)_T$ may be expressed as powers of $|\rho - \rho_c|$:

$$c_V \sim |\rho - \rho_c|^{-\alpha/\beta} \tag{1.2.28}$$

$$(\partial\rho/\partial p)_T \sim |\rho - \rho_c|^{1-\delta} \tag{1.2.29}$$

where β and δ are critical exponents with the values 0.33 and 4.82 respectively [20].

The effect of these divergences on the speed of sound can be deduced directly from equation (1.2.3) because $(\partial p/\partial T)_\rho$ does not show any critical 'anomalies'. Consequently, both terms in the equation vanish at the critical point. This result is not consistent with any analytic equation of state since all such equations erroneously predict a finite maximum in c_V at the critical point. Since α is considerably smaller than γ, it is, however, the behaviour of c_V that dominates along the coexistence curve and on the critical isochore for temperatures close to T_c. Similarly, α/β is small compared with $\delta - 1$ so that along the critical isotherm also the slow divergence of c_V dominates over the strong divergence of $(\partial\rho/\partial p)_T$ near to the critical point.

In practice several complications prevent a zero of the sound speed being obtained at the critical point of a pure fluid; instead a more or less sharp minimum is observed [21, 22]. The two main features complicating the situation are gravity-induced density gradients and critical-point dispersion. The former is just a reflection of the fact that the compressibility diverges so that even the weak pressure gradient induced in the fluid by gravity can have a significant effect on the density close to the critical state. To counter this effect, measurements have been made using apparatus in which the sample height is kept as small as possible [21]. Critical-point dispersion is associated with structural relaxation in the fluid on the scale of the correlation length,[5] which diverges in the approach to the critical point, and causes the sound speed measured using non-zero frequencies to differ from the 'thermodynamic' or zero-frequency value. There is a well developed theory, confirmed by experiment, to explain this [21]. Attempts to realize the thermodynamic limit have led to the use of low frequencies where relaxation effects become important only within a few millikelvin of the critical temperature [22].

A rich spectrum of critical phenomena is also observed in mixtures. The scaling laws can be generalized to describe, for example, those mixtures that separate into two phases above or below a critical temperature. The theoretical predictions, again supported by experiment, indicate that, while the isothermal compressibility diverges in the approach to the critical point along a constant composition path, c_V remains finite at the critical point and there is no strong 'anomaly' in u [23].

1.3   Kinetic Information from the Speed and Absorption of Sound

Unlike the low-frequency sound speed, which is determined by equilibrium properties, the absorption of sound is a consequence of kinetic processes in the fluid. These include the transport of heat and momentum, vibrational and rotational relaxation, and chemical or structural rearrangements taking place on a molecular scale. At high enough frequencies, the sound speed too becomes sensitive to these mechanisms and the absorption is accompanied by dispersion; both can provide some quantitative information about the timescale of various relaxation processes [24]. Since these effects will be examined in some detail in a later chapter, we confine the discussion here mainly to an outline of the kind of information available from acoustic measurements.

1.3.1   The absorption coefficient

We shall refer to just a single measure of absorption, namely the absorption coefficient α. This is defined such that the amplitude of a plane wave decays with distance z in proportion to $\exp(-\alpha z)$. All of the mechanisms that contribute to the overall absorption coefficient arise from the fact that sound waves disturb the local state of the fluid. Sound consists of longitudinal compression waves and it therefore imposes cyclic fluctuations in pressure and density which are generally associated with fluctuations in the local temperature. For the small amplitudes with which we will be concerned, these fluctuations are many orders of magnitude smaller than the equilibrium values of those quantities.

In the approach to zero frequency, where all processes are taking place slowly and the wavelength is long, the compressions and expansions of the fluid associated with the passage of sound occur reversibly and adiabatically; all of the sound energy is then propagated and none is dissipated. However, as the frequency is increased the fluid cannot continue to maintain local equilibrium on a microscopic scale because of the non-zero time required for the exchange of energy within and between molecules in response to fluctuations in the local density. The primary effect of this sloth is a failure of the acoustic cycle to occur reversibly. Consequently, some of the sound energy is dissipated and the absorption coefficient acquires a non-zero value. In addition, as the frequency is increased, the wavelength gets shorter and the temperature and velocity gradients in the fluid caused by the passage of sound become larger. Consequently, there is an increased flow of heat and momentum over distances comparable with the wavelength, the primary effect of which is again dissipation. Ultimately, all these processes affect also the sound speed but there is often a wide frequency range within which u is sensibly constant and α rises smoothly with frequency. Only when the period of the sound wave becomes comparable with a timescale characterizing one of the kinetic mechanisms do more dramatic effects occur.

1.3.2 Classical absorption

Amongst the various mechanisms mentioned above, the so-called classical absorption associated with the flows of heat and momentum is one that is always present. The magnitude of these effects is controlled by the transport coefficients $\varkappa$ and η, the coefficients of thermal conductivity and shear viscosity respectively. These give rise to a contribution to the absorption coefficient which we denote by α_{cl} and which is given by

$$\alpha_{cl} = (\omega^2/2u^3)[(4D_s/3) + (\gamma - 1)D_h] \tag{1.3.1}$$

where ω is the angular frequency, $\gamma = c_p/c_V$, and $D_s = \eta/\rho$ and $D_h = \varkappa/\rho c_p$ are the viscous and thermal diffusivities. This shows that the absorption increases in proportion to the square of the frequency. The magnitude of this term may be typified by the values $1.95 \times 10^{-3}\,\mathrm{m}^{-1}$ for argon gas at 300 K, 0.1 MPa and 10 kHz, and $6.7 \times 10^{-3}\,\mathrm{m}^{-1}$ for liquid water at the same temperature and pressure but at a frequency of 1 MHz (calculated using data from [25]). In principle, measurement of the absorption coefficient provides a route to the combination of the viscous and thermal diffusivities given by equation (1.3.1). However, other dissipation mechanisms are usually significantly greater than the classical viscothermal ones and this is not an attractive route. More usually, α_{cl} will be regarded as a small contribution to the total absorption which must be accounted for before other, more interesting, mechanisms can be quantified.

1.3.3 The bulk viscosity

At low enough frequencies all of the other relaxation mechanisms can be formally accounted for by the introduction of a frequency-independent bulk viscosity η_b which describes dissipation proportional to the local rate of compression. When this term is included, the total absorption coefficient is given by

$$\alpha = (\omega^2/2u^3)\,[(4D_s/3) + (\gamma - 1)D_h + (\eta_b/\rho)]\,. \qquad (1.3.2)$$

In a number of cases, the 'excess' absorption $(\alpha - \alpha_{cl})$ is dominated by one process (often vibrational relaxation) characterized by a single relaxation time that can be obtained from a measurement of the bulk viscosity. In other cases, either several mechanisms contribute to η_b or there may be one mechanism characterized not by a single relaxation time but by a spectrum of relaxation times.

1.3.4 Thermal relaxation

One of the most important processes leading to absorption and dispersion of sound in fluids, especially in gases, is the exchange of energy between translations and internal degrees of freedom of the molecules. The flow of energy into and out of storage in the vibrational and rotational modes of the molecules is driven by the fluctuations in the local temperature caused by sound; hence the term 'thermal relaxation' to describe time lags in this process. Usually the vibrational modes communicate with translations on a much slower timescale than do rotations and the process may be characterized by just a single overall relaxation time τ_{vib}. At frequencies small compared with τ_{vib}^{-1}, there is no dispersion but the relaxation process often dominates η_b and a precise measure of τ_{vib} may be obtained from that quantity. At higher frequencies, α increases dramatically, passes through a maximum at a frequency close to the inverse of the relaxation time, and then drops off again. In the same range, the sound speed increases, passes through a point of inflexion at a frequency also closely related to τ_{vib}^{-1}, and levels off at higher frequencies still. The magnitude of the absorption and dispersion is determined by that of the heat capacity associated with the mode or modes undergoing relaxation while the limiting behaviour at high frequencies is a consequence of that heat capacity 'dropping out' of the acoustic cycle; the internal mode or modes are ultimately unable to adjust significantly during the period of the sound wave and therefore no longer make a contribution to the effective heat capacity. Both u and the product $\alpha\lambda$, where λ is the wavelength, are illustrated as functions of frequency across the dispersive range in figure 1.3.

The situation with rotational relaxation is in principle the same, but whereas the complete behaviour illustrated in figure 1.3 can often be observed for vibrations, rotational relaxation is usually so fast that the frequencies at which substantial dispersion can occur are very high and translational effects are then of at least comparable magnitude. Nevertheless, much of our knowledge of rotational as well as vibrational relaxation times is acoustic in origin [26].

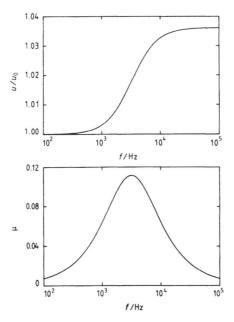

Figure 1.3 Sound speed u and absorption per wavelength μ as functions of frequency for carbon dioxide at 300 K and 0.1 MPa (u_0 is the zero-frequency speed of sound). Calculated using data given in [26].

1.3.5 Structural and chemical relaxation

Vibrational and rotational relaxations are but two kinds of molecular process activated by the passage of sound. The position of equilibrium between two or more chemical species present in the fluid can be displaced by the fluctuations of temperature and pressure associated with the sound. Where such displacement occurs reversibly, the effect simply contributes to the heat capacity and compressibility and there is no dispersion or excess absorption. At higher frequencies, comparable with the inverse of the time constant characterizing the rate of approach to equilibrium, absorption and dispersion occur leading to much the same kind of behaviour as illustrated above for vibrational relaxation. In a reaction between known and otherwise non-relaxing species, the sound speed is in principle determined by the position of equilibrium and the absorption coefficient by the rate constant. For simple reactions, the frequency at which the absorption is a maximum can be used to infer the rate constant in a known or postulated rate equation if the equilibrium constant is known. The magnitude of the absorption and dispersion is determined by the enthalpy and volume changes associated with the reaction.

Relaxation associated with structural rearrangement in the fluid can also be explained using a formalism based on reactions with specified rate constants and free-energy changes. These are phenomena exhibited by dense fluids and especially by those with specific interactions, such as hydrogen bonding, between molecules. They may or may not be characterized by one or a small number of relaxation times. For example, an explanation of critical-point absorption and dispersion involves a single relaxation time that varies with the distance from the critical point.

1.4　Transport Properties from Sound Absorption at a Surface

We have already seen that viscosity and thermal conduction are responsible for absorption of sound in the bulk of the fluid. However, the effects there are usually quite small because the velocity and temperature vary only over the relatively long distance of a wavelength and the corresponding gradients are therefore small. Near to a boundary with a solid surface this is no longer the case and much greater dissipation occurs—sufficient to form a basis for a measurement of the transport coefficients.

1.4.1　The boundary layers

To a rather good approximation, the boundary conditions that the sound field must satisfy require that the temperature fluctuations and tangential fluid velocity both vanish at the interface with a solid boundary despite their presence in the bulk of the fluid. This results in the establishment of boundary layers across which the temperature fluctuations and tangential fluid velocity rise from zero at the surface to their values in the bulk of the fluid. The thermal boundary layer may be characterized by a thickness, or penetration length, δ_h given by

$$\delta_h = (2D_h/\omega)^{1/2} = (2\kappa/\rho c_p \omega)^{1/2} \tag{1.4.1}$$

while the penetration length δ_s of the viscous boundary layer is

$$\delta_s = (2D_s/\omega)^{1/2} = (2\eta/\rho\omega)^{1/2}. \tag{1.4.2}$$

Since both these quantities are small compared with the wavelength λ, the gradients of temperature and tangential velocity are much greater than those in the bulk of the fluid and the irreversible flows of heat and momentum are correspondingly larger. Sound waves reflecting from the boundary with a solid surface therefore suffer a significant attenuation. When the incident sound wave is normal to the surface there is no tangential flow anyway and only thermal damping occurs; viscous damping is a maximum when the sound waves are travelling parallel to the surface.

1.4.2 Surface losses for an acoustic resonator

The most promising means by which the boundary layers can be exploited in an acoustic determination of viscosity and thermal conductivity is by the use of an acoustic resonator. Here a cavity of simple geometry, say a cylinder, is filled with the fluid under study and excited by a source of sound. When the frequency of the source is properly adjusted, a resonant standing wave can be established, the amplitude of which is controlled by the dissipation in the system. Under favourable conditions, the steady-state resonance linewidth, or the decay time for free oscillation after the source is shut off, provides a good measure of the surface damping and, through the relations which we will develop later, of the ratios δ_h/λ and δ_s/λ. In fact, for a given symmetry of standing wave, a linear combination of the two penetration lengths is obtained and, by using resonant modes of differing symmetry, it is possible to obtain values of both quantities. For example, the longitudinal and radial modes of a cylindrical resonator can be used to good effect [27]. The viscosity and thermal conductivity can then be determined if the density and isobaric specific heat capacity are known. For such measurements to be sensitive, other dissipation mechanisms must be kept small. Since the contribution of bulk absorption to a resonance linewidth increases in proportion to ω^2 while that of the boundary layer losses grows only in proportion to $\omega^{1/2}$, favourable conditions can be achieved at low frequencies. Although methods based on these principles have not yet been widely exploited, there is good reason to expect results that compare favourably with other techniques for the measurement of transport coefficients. The acoustic measurements which have been performed support this contention [27]. The theory also indicates that acoustic methods, unlike conventional techniques for measuring viscosity and thermal conductivity, are not sensitive to the slip and temperature-jump effects which arise at low pressures.[6] Thus the complication of estimating or measuring the jump coefficients that appear in the usual description of these phenomena can be avoided in methods that exploit sound absorption at a surface.

References

[1] Colclough A R, Quinn T J and Chandler T R D 1979 *Proc. R. Soc.* A **368** 125

[2] Moldover M R, Trusler J P M, Edwards T J, Mehl J B and Davis R S 1988 *Phys. Rev. Lett.* **60** 249; 1988 *J. Res. NBS* **93** 85

[3] Plumb H and Cataland G 1966 *Metrologia* **2** 18

[4] Colclough A R 1979 *Proc. R. Soc.* A **365** 349

[5] Moldover M R and Trusler J P M 1988 *Metrologia* **25** 165

[6] Ewing M B, Goodwin A R H, McGlashan M L and Trusler J P M 1987

J. Chem. Thermodyn. **19** 721 and 1988 *J. Chem. Thermodyn.* **20** 243; Ewing M B, Goodwin A R H and Trusler J P M 1989 *J. Chem. Thermodyn.* **21** 867

[7] Mehl J B and Moldover M R 1981 *J. Chem. Phys.* **74** 4062

[8] van Itterbeek A and Nihoul J 1955 *Acustica* **5** 142

[9] Colgate S O, Sona C F, Reed K R and Sivaraman A 1990 *J. Chem. Eng. Data* **35** 1

[10] van Dael W 1975 *Experimental Thermodynamics, Volume II: Experimental Thermodynamics of Non-reacting Fluids* ed. B Le Neindre and B Vodar (London: Butterworths) p 554

[11] Gammon B E 1976 *J. Chem. Phys.* **64** 2556

[12] van Dael W 1975 *Experimental Thermodynamics, Volume II: Experimental Thermodynamics of Non-reacting Fluids* ed. B Le Neindre and B Vodar (London: Butterworths) p 542

[13] Ewing M B, McGlashan M L and Trusler J P M 1987 *Mol. Phys.* **60** 681

[14] Ewing M B and Trusler J P M 1989 *J. Chem. Phys.* **90** 1106

[15] Kortbeek P J, Muringer M J P, Trappeniers N J and Biswas S N 1985 *Rev. Sci. Instrum.* **56** 1269

[16] Davis L A and Gordon R B 1967 *J. Chem. Phys.* **46** 2650

[17] Sun T F, Kortbeek, P J, Trappeniers N J and Biswas S N 1987 *Phys. Chem. Liq.* **16** 163

[18] Del Grosso V A and Mader C W 1972 *J. Acoust. Soc. Am.* **52** 1442

[19] Levelt Sengers J M H 1975 *Experimental Thermodynamics, Volume II: Experimental Thermodynamics of Non-reacting Fluids* ed. B Le Neindre and B Vodar (London: Butterworths) ch. 14

[20] Albert D Z 1982 *Phys. Rev.* B **25** 4810

[21] Thoen J and Garland C W 1974 *Phys. Rev.* A **10** 1331

[22] Garland C W and Williams R D 1974 *Phys. Rev.* A **10** 1328

[23] van Dael W 1975 *Experimental Thermodynamics, Volume II: Experimental Thermodynamics of Non-reacting Fluids* ed. B Le Neindre and B Vodar (London: Butterworths) pp 565–8

[24] Herzfeld K F and Litovitz T A 1959 *Absorption and Dispersion of Ultrasonic Waves* (London: Academic Press)

[25] Kaye G W C and Laby T H 1973 *Tables of Physical and Chemical Constants* 14th edn (London: Longman)

[26] Lambert J D 1977 *Vibrational and Rotational Relaxation in Gases* (Oxford: Clarendon)

[27] Carey C, Bradshaw J, Lin E and Carnevale E H 1974 *Experimental Determination of Gas Properties at High Temperatures and/or High Pressures* (Arnold Engineering Development Center, Arnold Air Force Station, TN 37389, USA) *Report no.* AEDC-TR-74-33, also available as NTIS AD-779772

[28] Dymond J H and Smith E B 1980 *The Virial Coefficients of Pure Gases and Mixtures* (Oxford: Clarendon) p 1

[29] Ewing M B, Owusu A A and Trusler J P M 1989 *Physica* A **156** 899

General references

Bett K E, Rowlinson J S and Saville G 1975 *Thermodynamics for Chemical Engineers* (London: Athlone)

Bhatia A B 1967 *Ultrasonic Absorption* (Oxford: Oxford University Press); reprinted by Dover Publications, New York, 1985

Cottrell T L and McCoubrey J C 1961 *Molecular Energy Transfer in Gases* (London: Butterworths)

Herzfeld K F and Litovitz T A 1959 *Absorption and Dispersion of Ultrasonic Waves* (London: Academic Press)

Le Neindre B and Vodar B (eds) 1975 *Experimental Thermodynamics, Volume II: Experimental Thermodynamics of Non-reacting Fluids* (London: Butterworths)

Maitland G C, Rigby M, Smith E B and Wakeham W A 1981 *Intermolecular Forces* (Oxford: Clarendon)

Matherson A J 1971 *Molecular Acoustics* (London: Wiley-Interscience)

McGlashan M L 1979 *Chemical Thermodynamics* (London: Academic Press)

Prausnitz J M, Lichtenlhaler R N and de Azevedo E G 1986 *Molecular Theory of Fluid Phase Equilibria* (Englewood Cliffs, NJ: Prentice-Hall)

Rowlinson J S and Swinton F L 1982 *Liquids and Liquid Mixtures* 3rd edn (London: Butterworths)

Woods L C 1975 *The Thermodynamics of Fluid Systems* (Oxford: Clarendon)

Notes

1. Nevertheless, familiarity with thermodynamic quantities and manipulations is assumed; some useful general references are given above.

2. *Isentropic* means 'at constant entropy'. Similarly, *isochoric* means 'at constant volume', *isobaric* 'at constant pressure' and *isothermal* 'at constant temperature'.

3. Quantum corrections, and ultimately a fully quantum mechanical treatment, may be necessary for light molecules (principally, H_2 and He) at low temperatures.

4. The speed of sound in sea water is also affected to a lesser extent by pressure and salinity.

5. The correlation length ζ is a distance over which fluctuations in the thermodynamic state of the fluid are correlated. Far from the critical point, ζ is of the order of the mean free path but it diverges like $|T - T_c|^{-0.63}$ in the approach to T_c along the critical isochore. Near to the critical point of a pure fluid, ζ can become comparable with the wavelength of light (say, 0.5 μm).

6. Slip and the temperature jump will be described in Chapter 2 (§2.4.3).

2

Fundamental Theory

2.1 Introduction

It is the task of this chapter to lay the theoretical foundations of the remainder of the book. Our emphasis is placed on relating measurable acoustic quantities, principally the speed and absorption of sound, to thermophysical properties of the medium. It is theory of that kind that allows acoustic measurements to serve as a tool in the study of thermodynamic and transport properties of gases and liquids. We shall devote most of our attention in this chapter to theory in which the medium is regarded as a classical continuum fluid. The results of this treatment will be applicable at frequencies small compared with those of molecular collisions or, in other terms, for wavelengths large compared with the mean free path. From the perspective of thermophysical property measurement this places few constraints upon the application of acoustic techniques; in practice, the only important effects arising from the finite length of the mean free path occur for dilute gases in the boundary layers near to an interface. Such small effects may be treated with sufficient accuracy by very simple kinetic theory. Molecular energy transfer and other relaxation mechanisms can be of much greater importance because they may be characterized by relaxation times orders of magnitude greater than the time between binary collisions. Here we include in our treatment of dissipation a bulk viscosity that arises from relaxation mechanisms other than the classical viscothermal processes, but we delay to a later chapter discussion both of the mechanisms themselves and of their consequences at frequencies that are not small compared with the inverse of the relevant relaxation time. Of course dynamic effects at high frequencies are of great interest in themselves. Attempts to explain the propagation of sound in rarefied gases at frequencies comparable with the molecular collision frequency touch on fundamental limitations within our present kinetic theory, while evaluation of the bulk viscosity from molecular properties presents severe difficulties. However, these are effects beyond the scope of this chapter.

22

2.2 Propagation in an Idealized Fluid

In many of the problems considered in this book the primary quantities of interest are equilibrium properties of the fluid. The propagation of sound is strongly influenced by thermodynamic properties and, it turns out, usually only weakly affected by non-equilibrium phenomena. Although we ultimately seek a theory that adequately describes both propagation and dissipation of acoustic energy, we first derive a simplified theory of sound that neglects completely the effects of mechanisms that result in dissipation. Thus the whole of this section is strictly applicable only to a fluid in which viscosity and thermal conduction are negligible and in which local thermodynamic equilibrium is achieved instantaneously. Should the fluid comprise two or more components then we further assume that the composition remains entirely homogeneous; we neglect both chemical reactions and diffusion processes. Such a fluid is called ideal. In the course of this preliminary encounter with fluid mechanics we introduce several important elements of the more complete theory and, in particular, derive an expression for the speed of sound that turns out to be of high accuracy.

2.2.1 The equation of continuity

In the following discussion we adopt the Euler description of the fluid in which the frame of reference is fixed in space and a property of the fluid, P for example, refers to the portion of fluid that happens to be at the position r of interest at the time t. The equation of continuity is a mathematical statement of the obvious: if P is an extensive property then any change in the amount of P within a closed surface must be caused either by fluid flow or by creation of P within that region. We consider some volume V_0 in space and denote the density of P by P'. The amount of P flowing in unit time through an element dS of the surface that bounds V_0 is $n \cdot J \, dS$, where the vector $J = P'v$ is the current density of P, n is the outward-pointing unit vector normal to the surface and v is the fluid velocity. Thus $n \cdot J \, dS$ is positive for outflow and negative for inflow. Clearly, the rate of outflow from the volume V_0 is given by the integral $\iint n \cdot J \, dS$ performed over the whole of the surface that bounds V_0. This surface integral may be transformed to an integral over the volume using the divergence theorem: thus

$$\iint n \cdot J \, dS = \iiint \nabla \cdot J \, dV \qquad (2.2.1)$$

where $\nabla \cdot J$ is the divergence of J.[1] If in addition P is created at a rate $H(r, t)$ per unit volume then the net rate at which P accumulates within V_0 must be $\iiint (H - \nabla \cdot J) \, dV$. Since the total amount of P within the volume is $\iiint P' \, dV$, the rate of change in the density of P must be given by the integral equation

$$\iiint [(\partial P'/\partial t) + \nabla \cdot J - H] \, dV = 0 \qquad (2.2.2)$$

and, since this expression must be true of any volume, the integrand vanishes and we obtain

$$(\partial P'/\partial t) + \nabla \cdot J = H. \qquad (2.2.3)$$

Equation (2.2.3) is called the equation of continuity for the density of P. As yet, we have introduced no approximations.

2.2.2 Euler's equation

We now derive the fundamental equation of motion for an ideal fluid by application of Newton's second law of motion to an element of volume dV. In order to evaluate the force acting on such a volume, we first consider a cubic element $dx \, dy \, dz$ with edges parallel to the Cartesian axes. This situation is illustrated in figure 2.1. The force acting in the x direction across the face at x is $p(x) \, dy \, dz$, where $p(x)$ is the pressure there, while the force acting across the face at $x + dx$ is $[p(x) + (\partial p/\partial x) \, dx] \, dy \, dz$. Consequently, the x component of the net force acting on the volume element is just the difference between these two values: $-(\partial p/\partial x) \, dx \, dy \, dz$. Since analogous arguments apply along the other two axes, the total net force acting on the element is $-\nabla p \, dx \, dy \, dz$ and we can identify $-\nabla p$ as the force per unit volume acting at any point in the fluid. The force acting on an element of arbitrary shape is therefore $-\nabla p \, dV$ and, since the mass is $\rho \, dV$ (where ρ is the mass density), Newton's law requires that

$$-\nabla p = \rho(dv/dt). \qquad (2.2.4)$$

The total derivative (dv/dt) that appears in this equation refers not to a fixed location in space but to a moving portion of the fluid; it is related to the partial derivative $(\partial v/\partial t)$ at some fixed location r by[2]

$$(dv/dt) = (\partial v/\partial t) + (v \cdot \nabla)v. \qquad (2.2.5)$$

It follows that the equation of motion of an ideal fluid is

$$-\nabla p = \rho(\partial v/\partial t) + \rho(v \cdot \nabla)v. \qquad (2.2.6)$$

This is the equation first derived by Euler. Since we have neglected friction, the equations of this section and those that we shall derive from them are restricted to an ideal fluid. Our description of the fluid now contains five unknowns: the three components of v, the mass density and the pressure. Five equations are therefore required to specify a solution and for this purpose we thus far have four: the equation of continuity for mass density and Euler's equation (really three equations in the components of v). Finally we note that for our ideal fluid (in which there is no friction or thermal conduction) all motion must be adiabatic and reversible; in other words isentropic. Thus, denoting entropy by S, our fifth equation relates a small change δp in the pressure to the small change $\delta\rho$ in the mass density by

$$\delta p = (\partial p/\partial \rho)_S \delta\rho. \tag{2.2.7}$$

Since the local thermodynamic state of the ideal fluid is completely specified by the values of the two independent quantities p and ρ, and the motion is specified by the three components of v, these five quantities are sufficient to determine the state of the moving fluid.

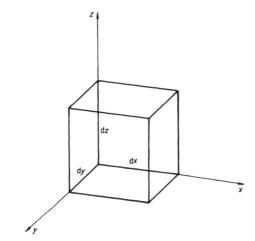

Figure 2.1 A cubic element of volume in the fluid.

2.2.3 The linearized wave equation

It is now possible to derive an equation to describe the propagation of sound for the limiting case in which the amplitude of the disturbance is small and dissipative mechanisms are negligible. In the presence of sound we denote the pressure exerted by the fluid by $p + p_a(r, t)$, where p is the equilibrium pressure, assumed constant throughout the fluid, and $p_a(r, t)$ is the small additional term arising from the sound field; we call this the *acoustic pressure*. Similarly, the temperature will be denoted by $T + T_a(r, t)$ and the mass density by $\rho + \rho_a(r, t)$, where $T_a(r, t)$ and $\rho_a(r, t)$ are the acoustic contributions to the temperature and density. Since the fluid is assumed at rest except for the motion due to the passage of sound, the fluid velocity v is also one of the 'small' acoustic quantities. The essential assumptions of linear acoustics are that the acoustic contributions to the pressure, temperature and density are small compared with the equilibrium values, and that the fluid speed is small compared with the sound speed; with these assumptions we neglect squares, cross-products and higher powers of the small acoustic quantities. Here we assume also that in the presence of sound *local* thermodynamic equilibrium is established instantaneously in the fluid.

In terms of our new notation, the equation of continuity for mass density becomes

$$(\partial \rho_a / \partial t) = -(\rho + \rho_a) \nabla \cdot v - (v \cdot \nabla) \rho_a \qquad (2.2.8)$$

and Euler's equation may be written

$$-\nabla p_a = (\rho + \rho_a)[(\partial v / \partial t) + (v \cdot \nabla)v]. \qquad (2.2.9)$$

Approximated to leading order in the small acoustic terms, these equations become

$$\rho^{-1}(\partial \rho / \partial p)_S (\partial p_a / \partial t) = -\nabla \cdot v \qquad (2.2.10)$$

$$\rho(\partial v / \partial t) = -\nabla p_a \qquad (2.2.11)$$

where we have eliminated $(\partial \rho_a / \partial t)$ in favour of $(\partial p_a / \partial t)$ using the isentropic equation of state, equation (2.2.7). We next take the divergence of equation (2.2.11) and, applying (2.2.10) to eliminate v, obtain

$$[\nabla^2 - (1/u_0^2)(\partial^2 / \partial t^2)] p_a(r, t) = 0 \qquad (2.2.12)$$

where $\nabla^2 = \nabla \cdot \nabla$ is the Laplacian operator and we have used the notation $u_0^2 = (\partial p / \partial \rho)_S$. Equation (2.2.12) is a wave equation and describes the propagation of a small-amplitude acoustic disturbance in a non-dissipative medium.

A general solution of the wave equation for any initial conditions may be obtained but this approach will not be pursued here [1]. A particular solution is given by $p_a = f(q)$ where $q = r + uth$; clearly, this function describes a wave travelling in a direction parallel to that of the unit vector h at speed u. On substitution of f into the wave equation we find that the speed of sound u is equal to $u_0 = (\partial p / \partial \rho)_S^{1/2}$; we call u_0 the ideal speed of sound. In this example the motion is completely homogeneous in a plane perpendicular to h. Such a wave is called a plane wave. Substitution in equation (2.2.11) shows that the direction of the fluid motion is identical with that of the unit vector h. Since the directions of motion and propagation are the same, sound waves are said to be longitudinal. Under many circumstances the speed of sound in a real fluid is very close to that of the corresponding ideal fluid. At 300 K and 0.1 MPa the speed of sound is 322 m s^{-1} in argon and 1501 m s^{-1} in water.

2.2.4 The velocity potential

Before proceeding to discuss the influence of dissipative mechanisms in the linearized equations of motion, it is useful to recast the simplified description of the wave field. In the absence of dissipation, the fluid flow will be potential or irrotational flow[3] and in consequence the fluid velocity may be expressed as the gradient of a scalar function. This function is called the velocity potential which we denote by $\Psi(r, t)$ so that

$$v(r, t) = -\nabla\Psi(r, t) \qquad (2.2.13)$$

and using equation (2.2.11), the linearization of Euler's equation, we have to first order in the acoustic variables

$$p_a(r, t) = \rho(\partial\Psi(r, t)/\partial t). \qquad (2.2.14)$$

Thus we may express both fluid velocity and the acoustic pressure simply in terms of the velocity potential Ψ. Since instantaneous local equilibrium is assumed, the other acoustic variables may also be expressed in terms of the velocity potential through standard thermodynamic manipulations.

2.3 Propagation in a Dissipative Fluid

In the previous sections we have introduced two classes of assumptions: those that are common to all of linear acoustics, namely that the acoustic variables are sufficiently small for their squares, cross-products and higher powers to be neglected; and those adopted to obtain the simple solution, namely the establishment of instantaneous local equilibrium in the fluid and the neglect of dissipation. The linear theory of sound is exact in the limit $p_a \to 0$ and in practice the small-amplitude limiting behaviour is easily achieved. In contrast, the additional approximations made above do not become exact in that limit and require further discussion.

In constructing a description of sound that rectifies the shortcomings of the simplified treatment we must take account of both thermal conduction and friction in the fluid; these are always present. In addition there may be other mechanisms that inhibit the establishment of local thermodynamic equilibrium. Although not always present to any appreciable degree, molecular and chemical relaxation processes can have very important effects; these will be discussed in a later chapter.

2.3.1 The Navier–Stokes equation

The effects of friction enter directly into the equations of motion. For simplicity, we again consider the fluid within a cubic element of volume

$dx\,dy\,dz$ around r (figure 2.1). The force acting across the face perpendicular to x is $\mathcal{F}_x(r, t)\,dy\,dz$, where the vector $\mathcal{F}_x$ represents the fluid stress on that face. In the absence of viscosity, $\mathcal{F}_x$ is very simple, having the direction of the x-pointing unit vector i and magnitude equal to the sum of the hydrostatic and acoustic pressures. However, in a viscous fluid the interaction between two regions of the fluid separated by a plane surface is no longer normal to that surface. Therefore $\mathcal{F}_x$ does not point exactly in the x direction but has three components:

$$\mathcal{F}_x = i\,\mathcal{F}_{xx} + j\,\mathcal{F}_{xy} + k\,\mathcal{F}_{xz} \qquad (2.3.1)$$

where i, j, k are the unit base vectors of the Cartesian coordinate system. The resultant force on the volume element transmitted across the two faces perpendicular to the x axis is $-(\partial\mathcal{F}_x/\partial x)\,dV$ and similar terms arise for the resultant forces acting across the other faces. The total net force acting on the element of volume is therefore

$$F(r, t) = -dV[(\partial\mathcal{F}_x/\partial x) + (\partial\mathcal{F}_y/\partial y) + (\partial\mathcal{F}_z/\partial z)] = -dV\,\nabla\cdot\mathcal{F} \quad (2.3.2)$$

and will contribute to the rate of change of momentum within the volume element. The second form of this equation is written in terms of the stress tensor $\mathcal{F}$ which may be given the matrix representation

$$\mathcal{F} = \begin{pmatrix} \mathcal{F}_{xx} & \mathcal{F}_{xy} & \mathcal{F}_{xz} \\ \mathcal{F}_{yx} & \mathcal{F}_{yy} & \mathcal{F}_{yz} \\ \mathcal{F}_{zx} & \mathcal{F}_{zy} & \mathcal{F}_{zz} \end{pmatrix}. \qquad (2.3.3)$$

The operation $\nabla\cdot\mathcal{F}$ is understood to result in the vector with q_j component $\Sigma_i\,(\partial\mathcal{F}_{ij}/\partial q_i)$, where q_i, q_j represent the Cartesian coordinates x, y or z in turn as i, j take values $1, 2, 3$. From the discussion above, it is clear that for a non-viscous fluid the elements $\mathcal{F}_{ij}$ of the stress tensor are given by the simple formula

$$\mathcal{F}_{ij} = (p + p_a)\delta_{ij} \qquad (2.3.4)$$

in which δ_{ij} is the Dirac delta function ($\delta_{ii} = 1$; $\delta_{ij} = 0$, $i \neq j$). In the presence of viscosity we denote the elements of the full stress tensor by

$$\mathcal{F}_{ij} = (p + p_a)\delta_{ij} + \mathcal{F}'_{ij} \qquad (2.3.5)$$

where $\mathcal{F}'$ is called the viscous stress tensor. To formulate this viscous stress tensor we must consider the diffusion of momentum that arises in the presence of gradients in the velocity. There can be no elements in $\mathcal{F}'$ independent of such gradients because internal friction is absent when the fluid is at rest or in uniform motion. Furthermore, there is no internal friction when the fluid is in uniform rotation about the frame of reference. Since there are many accounts in specialized texts on fluid mechanics (see, for example, [2]), we do not give here a complete derivation of the components of $\mathcal{T}'$. However, we note that it is convenient to resolve the internal friction into two terms, proportional to the rates of shear and of compression respectively, and that the result may be written

$$\mathcal{T}'_{ij} = -\eta\{[(\partial v_i/\partial q_j) + (\partial v_j/\partial q_i)] - \tfrac{2}{3}\delta_{ij}\nabla \cdot v\} - \eta_b(\nabla \cdot v)\delta_{ij} \quad (2.3.6)$$

where we have used the notation v_i to denote the q_i components of v. In equation (2.3.6), η and η_b are transport coefficients: η is called the coefficient of shear viscosity and is associated with shear motion in the fluid; η_b is the coefficient by bulk viscosity[4] and has its origins in relaxation phenomena associated with compression of the fluid. The bulk viscosity will be considered in some detail in Chapter 4.

The Navier–Stokes equation for a compressible fluid is obtained using the full stress tensor $\mathcal{T}$ in the expression for the force acting on a volume element: since the mass of fluid within this element is $(\rho + \rho_a)\,dV$ we have

$$-\nabla \cdot \mathcal{T} = (\rho + \rho_a)(dv/dt). \quad (2.3.7)$$

This relation is now used in place of Euler's equation. The problem may be simplified by separating $\mathcal{T}$ into the sum

$$\mathcal{T} = [p + p_a - (\eta_b + 4\eta/3)(\nabla \cdot v)]\mathcal{I} + \mathcal{J} \quad (2.3.8)$$

in which $\mathcal{I}$ is the nine-component symmetric tensor with elements δ_{ij} and $\mathcal{J}$ has components

$$\mathcal{J}_{ij} = \eta\{2(\nabla \cdot v)\delta_{ij} - [(\partial v_i/\partial q_j) + (\partial v_j/\partial q_i)]\}. \quad (2.3.9)$$

Now, since it can be shown that

$$-\nabla \cdot \mathcal{J} = -\eta\nabla \times (\nabla \times v) \quad (2.3.10)$$

where the vector product $\nabla \times v$ is the curl of v, equation (2.3.7) may be written correct to first order in the acoustic variables in the form

$$(\partial v/\partial t) = -(1/\rho)\nabla p_a + [(4D_s/3) + (\eta_b/\rho)]\nabla(\nabla \cdot v) - D_s\nabla \times (\nabla \times v)$$
$$(2.3.11)$$

where $D_s = \eta/\rho$ is called the viscous diffusivity. This separation is useful because, being a vector function of position, the fluid velocity v can be resolved into the sum of a longitudinal (or irrotational) component v_l for which $\nabla \times v_l = 0$, and a rotational (or transverse) component v_r for which $\nabla \cdot v_r = 0$. Thus the longitudinal component may be represented in terms of a velocity potential. In addition, the gradient of a scalar function is entirely irrotational and therefore ∇p_a contributes only to the longitudinal fluid flow. Consequently the linearized Navier–Stokes equation, equation (2.3.11), may be written as two uncoupled equations:

$$(\partial v_l/\partial t) = -(1/\rho)\nabla p_a + D_v\nabla(\nabla \cdot v_l) \quad (2.3.12)$$

and[5]

$$(\partial v_r/\partial t) = -D_s \nabla \times (\nabla \times v_r) = D_s \nabla^2 v_r \qquad (2.3.13)$$

where, to simplify the notation, we have introduced the quantity $D_v = (4D_s/3) + (\eta_b/\rho)$ by analogy with the viscous diffusivity. Since equation (2.3.13) is unrelated to the acoustic pressure it may be neglected in the bulk of the fluid. However, transverse flow may be important when there are boundary conditions to be satisfied.

2.3.2 Thermal conduction

The fluctuations in pressure characteristic of sound are of course accompanied by fluctuations in the temperature. Thus gradients in the temperature will exist in the presence of sound and irreversible heat flow from regions of higher temperature to those of lower temperature will occur. In consequence, the acoustic cycle fails to be perfectly isentropic.

According to Fourier's equation, the heat current density J_h is proportional to the gradient of the temperature:

$$J_h = -\varkappa \nabla T_a. \qquad (2.3.14)$$

The constant of proportionality $\varkappa$ is called the coefficient of thermal conductivity. Denoting the heat per unit mass by q_a, and neglecting frictional source terms (which are of the second order), we next apply the equation of continuity to determine the rate at which the acoustic heat density ρq_a changes with time. To a consistent first-order approximation we find:

$$\begin{aligned} \rho(\partial q_a/\partial t) &= -\nabla \cdot J_h \\ &= \varkappa \nabla^2 T_a. \end{aligned} \qquad (2.3.15)$$

The rate of entropy production per unit mass may now be obtained by invoking the relation

$$ds = dq/T \qquad (2.3.16)$$

which applies to a reversible change and is therefore correct here to a first-order approximation. Combining this expression with equation (2.3.15) we find

$$(\partial s_a/\partial t) = (\varkappa/\rho T)\nabla^2 T_a \qquad (2.3.17)$$

in which s_a is the small acoustic contribution to the specific entropy. s_a may be related to the acoustic temperature and pressure by the relation

$$s_a = (\partial s/\partial T)_p T_a + (\partial s/\partial p)_T p_a \qquad (2.3.18)$$

and, since $(\partial s/\partial T)_p = c_p/T$ and $(\partial s/\partial p)_T = (1/\rho^2)(\partial \rho/\partial T)_p$, the rate of entropy production per unit mass is also given by

$$(\partial s_a/\partial t) = (c_p/T)(\partial T_a/\partial t) + (1/\rho^2)(\partial\rho/\partial T)_p(\partial p_a/\partial t). \qquad (2.3.19)$$

Finally, eliminating the entropy between equations (2.3.17) and (2.3.19), and making use of the identity $(T/c_p\rho^2)(\partial\rho/\partial T)_p = -(\gamma-1)/\gamma\beta$, where $\beta = (\partial p/\partial T)_\rho$, we obtain the second-order diffusion equation:

$$D_h\nabla^2 T_a = (\partial/\partial t)\{T_a - [(\gamma-1)/\gamma\beta]\, p_a\}. \qquad (2.3.20)$$

The transport coefficient $D_h = \varkappa/\rho c_p$ that appears in this equation is called the thermal diffusivity.

2.3.3 The modified wave equation

We first restrict our attention to the longitudinal terms in the hydrodynamic equations. Our description now contains six quantities: the acoustic pressure p_a, temperature T_a, density ρ_a, and the three components of the fluid velocity v. Six equations are required to specify a solution.

The first three of these equations are the three components of the linearized Navier–Stokes equation for the irrotational flow of a compressible fluid, equation (2.3.12). The fourth expression in the equation of mass–density continuity, which we now write as

$$(\partial\rho_a/\partial t) + \rho\nabla\cdot v_l = 0 \qquad (2.3.21)$$

and the fifth is the heat diffusion equation, equation (2.3.20). Finally, we use the thermodynamic equation of state to interrelate the acoustic contributions to the density, pressure and temperature:

$$\rho_a = (\partial\rho/\partial p)_T p_a + (\partial\rho/\partial T)_p T_a$$
$$= (\gamma/u_0^2)(p_a - \beta T_a). \qquad (2.3.22)$$

Here we have used the relations $(\partial\rho/\partial p)_T = \gamma(\partial\rho/\partial p)_s$, $(\partial\rho/\partial T)_p = -(\partial p/\partial T)_\rho(\partial p/\partial\rho)_T$ and $u_0^2 = (\partial p/\partial\rho)_s$.

To obtain the modified wave equation we first take the divergence of equation (2.3.12); we next eliminate v_l using the equation of mass–density continuity to obtain

$$\nabla^2 p_a = (\partial^2\rho_a/\partial t^2) - D_v\nabla^2(\partial\rho_a/\partial t). \qquad (2.3.23)$$

Finally ρ_a is eliminated using (2.3.22) with the result

$$\nabla^2 p_a = (\gamma/u_0^2)[(\partial^2/\partial t^2) - D_v(\partial/\partial t)\nabla^2](p_a - \beta T_a). \qquad (2.3.24)$$

This equation may be solved simultaneously with equation (2.3.20) for T_a and p_a. We note that, should it be required, the longitudinal fluid velocity may be found from

$$(\partial v_l/\partial t) = -\nabla[(p_a/\rho) + (\gamma D_v/\rho u_0^2)(\partial/\partial t)(p_a - \beta T_a)] \qquad (2.3.25)$$

the combination of equations (2.3.12), (2.3.21) and (2.3.22) that eliminates ρ_a.

2.3.4   Simple-harmonic wave motion

We now investigate solutions of the modified wave equation. To obtain these solutions in a convenient form we consider the case of mono-frequency waves propagating in free space. In such an analysis all of the small acoustic variables have a harmonic dependence in time; the factor for such motion is exp($i\omega t$), where $\omega/2\pi$ is the frequency of the sound. Since *any* wave may be expanded into harmonic components of various frequencies, using a Fourier series or integral, this approach places no constraints whatsoever on the solutions that we shall obtain. The acoustic pressure will therefore be required to satisfy the equation

$$(\partial/\partial t)p_a = i\omega p_a. \tag{2.3.26}$$

Since the acoustic temperature and pressure must be proportional to each other in a monofrequency wave it follows from equations (2.3.24) and (2.3.26) that p_a must be an eigenfunction of the Laplacian operator. We therefore also require that p_a must be solution of the Helmholtz equation

$$(\nabla^2 + k^2)p_a = 0 \tag{2.3.27}$$

where k is called the propagation constant. This section is devoted to evaluating the eigenvalues $-k^2$ while in the following section the eigenfunctions appropriate to plane travelling waves will be examined. In order to simplify the notation, it is useful to introduce a dimensionless reduced propagation constant $\Gamma = ku_0/\omega$ defined such that Γ takes the value unity for an ideal fluid. Generally, we will find that $\mathrm{Re}(\Gamma) = (u_0/u)$ and $\mathrm{Im}(\Gamma) = \alpha u_0/\omega$, where u is the phase speed and α the coefficient of absorption at angular frequency ω.

Inserting $-\Gamma^2$ for $(u_0^2/\omega)\nabla^2$ and $i\omega$ for $(\partial/\partial t)$ into equation (2.3.24) we obtain

$$(\Gamma^2 + i\gamma\omega\tau_v\Gamma^2 - \gamma)p_a + (\gamma - i\gamma\omega\tau_v\Gamma^2)\beta T_a = 0 \tag{2.3.28}$$

where $\tau_v = D_v/u_0^2$. Similarly, from equation (2.3.20) we obtain the expression

$$(1 - i\omega\tau_h\Gamma^2)T_a - [(\gamma - 1)/\gamma\beta]\,p_a = 0 \tag{2.3.29}$$

in which $\tau_h = D_h/u_0^2$. Then combining equations (2.3.28) and (2.3.29), we obtain the following equation for Γ^2:

$$\Gamma^4(\gamma\omega^2\tau_v\tau_h - i\omega\tau_h) + \Gamma^2(1 + i\gamma\omega\tau_h + i\omega\tau_v) - 1 = 0. \tag{2.3.30}$$

This expression is again subject to the principal condition that the amplitude of the sound is small. However, full account has been taken of the effects of thermal conductivity and viscosity, so that equation (2.3.30) is essentially exact in the small-amplitude limit. It should be noted that this is a quadratic expression in Γ^2 with two roots and that, together with the solution of the transverse wave equation, there are now three valid solutions in place of the single solution of §2.2.

The exact solutions of equation (2.3.30) are given by

$$\Gamma^2 = \frac{-i}{2\omega\tau_h} \left(\frac{1 + i\omega\tau_v + i\gamma\omega\tau_h \pm D}{1 + i\gamma\omega\tau_v} \right) \qquad (2.3.31)$$

where D^2 may be written

$$D^2 = (1 + i\omega\tau_v - i\gamma\omega\tau_h)^2 + 4i(\gamma - 1)\omega\tau_h. \qquad (2.3.32)$$

Usually the effects of thermal conductivity and viscosity are slight and the quantities $\omega\tau_h$ and $\omega\tau_v$ correspondingly small. Under these conditions an approximation to the exact solution is appropriate. Expanding D in a binomial series and retaining terms of orders up to the first in $\omega\tau_v$ and the second in $\omega\tau_h$ we find[6]

$$D \approx 1 + 2(\gamma - 1)\omega\tau_h\omega(\tau_v - \tau_h) + i[\omega(\tau_v - \tau_h) + (\gamma - 1)\omega\tau_h]. \quad (2.3.33)$$

Then, taking $-D$ in equation (2.3.31), we obtain the first solution of the longitudinal wave equation:

$$\Gamma^2 = 1 - i[\omega\tau_v + (\gamma - 1)\omega\tau_h]. \qquad (2.3.34)$$

We call this the propagational mode of sound. Equation (2.3.34) shows that, in the first-order approximation, inclusion of the thermal conductivity and viscosity leaves the real part of the propagation constant unchanged; the wave speed is frequency independent and identical with the ideal speed of sound u_0. However, the dissipative mechanisms lead to a small imaginary component of Γ, proportional to ω, that describes energy loss from the wave. For gases the relaxation times $\tau_h = D_h/u_0^2$ and $\tau_s = D_s/u_0^2$ characterizing the classical processes of heat and momentum transfer are of the same order as the mean time τ_c between binary molecular collisions, so when the bulk viscosity is of the same order as the ordinary shear viscosity, the first-order treatment of dissipative mechanisms is accurate provided that $\omega\tau_c \ll 1$. For example, in argon at 0.1 MPa and 300 K, $\tau_h \approx 0.20$ ns and $\tau_s \approx 0.14$ ns. Even at the quite high frequency of 1 MHz the sound speed obtained from the exact solution, equation (2.3.31), does not differ from u_0 by more than a few parts in 10^6. In most liquids τ_h and τ_s are very small, typically being of order 10^{-12} s. For example, at 300 K in pure water, $\tau_h = 0.06$ ps and $\tau_s = 0.38$ ps. This means that in the absence of significant relaxation processes other than heat and momentum transfer a first-order treatment of dissipation is highly accurate under usual conditions.

Substitution of equation (2.3.34) into (2.3.29) and (2.3.25) shows that the propagational mode makes contributions to the acoustic temperature and the longitudinal fluid velocity given by

$$T_p = [(\gamma - 1)/\gamma\beta] \, (1 + i\omega\tau_h) p_p \qquad (2.3.35)$$

$$v_{l,p} = (i/\omega\rho)(1 + i\omega\tau_v)\nabla p_p \qquad (2.3.36)$$

where p_p is the acoustic pressure of the propagational mode.

The second solution of the longitudinal wave equation, corresponding to the root with $+D$ in equation (2.3.31), is given in first order by

$$\Gamma_h^2 = -i/(\omega\tau_h) \qquad (2.3.37)$$

and is called the thermal mode. Here, the reduced propagation constant Γ_h has equal real and imaginary parts indicating that thermal waves suffer very rapid attenuation. Substitution in equation (2.3.28) reveals that

$$p_h = -i\gamma\beta\omega(\tau_h - \tau_v)T_h \qquad (2.3.38)$$

is the contribution of the thermal mode to the acoustic pressure, from which it follows that the contribution to the longitudinal fluid velocity is

$$v_{l,h} = (\gamma\beta\tau_h/\rho)\nabla T_h. \qquad (2.3.39)$$

Although for a real fluid it will be necessary to include the thermal mode when there are boundary conditions to be satisfied, in the bulk of the fluid it can usually be neglected.

A third solution arises from the rotational terms in the hydrodynamic equations. This is called the shear mode and, like the thermal mode, can usually be ignored in the bulk of the fluid. Again, however, shear waves can be important in the region close to a boundary of the fluid. Imposing a harmonic solution on equation (2.3.13), the rotational part of the linearized Navier–Stokes equation, we find

$$\Gamma_s^2 = -i/(\omega\tau_s) \qquad (2.3.40)$$

where $\tau_s = D_s/u_0^2$ is independent of the bulk viscosity. There are no corresponding contributions to the acoustic temperature or pressure.

2.3.5 Plane simple-harmonic waves

To illustrate the effects of dissipation in the fluid, we consider the behaviour of plane monofrequency waves propagating in free space. Let the direction of propagation be that of the z axis. Then all of the acoustic variables are functions just of the variable $ut - z$. For the propagational mode, the acoustic pressure will be

$$p_p(z, t) = f(z)\exp(i\omega t) \qquad (2.3.41)$$

where $f(z)$ is an eigenfunction of the Laplacian operator with eigenvalue $-k^2$; two solutions are given by $f(z) = A \exp(\pm ikz)$, where A is constant. The positive-going wave is therefore

$$p_p = A \exp[i(\omega t - kz)] = A \exp(-\alpha z)\exp[i\omega t - (\omega/u_0)z] \quad (2.3.42)$$

where $\alpha = -\operatorname{Im}(k)$ is the coefficient of absorption and is given by

$$\begin{aligned}\alpha &= (\omega^2/2u_0)[\tau_v + (\gamma - 1)\tau_h]\\ &= (\omega^2/2u_0^3)[(4D_s/3) + (\gamma - 1)D_h + (\eta_b/\rho)]. \quad (2.3.43)\end{aligned}$$

In the more general case where the direction of propagation is that of the arbitrary unit vector h, we simply replace kz by $k \cdot r$ wher k is called the wavevector and has magnitude k and direction h.

In dilute monatomic gases the bulk viscosity should vanish and then the absorption coefficient is determined by the classical viscothermal mechanisms. For argon at 300 K and 0.1 MPa the classical absorption coefficient is given by $\alpha_{cl}/f^2 = 1.95 \times 10^{-11}$ s^2 m^{-1}. Experimental determinations of α show good agreement between theory and experiment for dilute monatomic gases except at very high frequencies where the hydrodynamic description is bound to fail [3]. In most other gases α exceeds α_{cl} because of the finite bulk viscosity; this is also true for most liquids. For example, in pure water α_{cl}/f^2 is calculated to be 6.7×10^{-15} s^2 m^{-1} but, although the expected frequency dependence is confirmed at frequencies up to 70 MHz, the experimental value is 21×10^{-15} s^2 m^{-1} [4].

The longitudinal component of the fluid velocity associated with the plane wave of equation (2.3.42) is, by definition, directed along the z axis; it is given by

$$v_l = [1 + \tfrac{1}{2}i\omega\tau_v - \tfrac{1}{2}(\gamma - 1)i\omega\tau_h] (p_p/\rho u_0). \quad (2.3.44)$$

This equation illustrates the important analogy between the quotient p_a/v in acoustics and electrical impedance in AC circuit theory. We see that for the plane wave in an ideal fluid $p_a/v = \rho u_0$ and for that reason ρu_0 is called the characteristic acoustic impedance of the medium. Inclusion of the small imaginary terms in equation (2.3.44) results in a slight phase difference between the acoustic pressure and the fluid velocity but, to a first-order approximation, no change of magnitude. Although it will frequently be necessary to consider the attenuation of sound in space or in time resulting from viscosity and heat conduction, for other purposes it is usually sufficient to neglect this small phase difference between p_a and v. Since the time average of the potential energy density is equal to $\tfrac{1}{2} \times s |p_a|^2$ and that of the kinetic energy density is $\tfrac{1}{2}\rho |v|^2$, the time average of the total acoustic energy density is equal to $w = |p_a|^2/\rho u^2$. The time-average intensity (power flux per unit area) uw is therefore given by

$$\mathfrak{I} = |p_a|^2/\rho u. \quad (2.3.45)$$

The wavefunction of a plane thermal wave must also be an eigenfunction of ∇^2 and in this case the eigenvalue is $-k_h^2$. A positive-going thermal wave is therefore given by

$$T_h = B \exp(i\omega t - ik_h z) = B \exp[i\omega t - (1 + i)z/\delta_h] \qquad (2.3.46)$$

where $k_h = (\omega/u_0)\Gamma_h$ and δ_h is a constant given by

$$\delta_h = (1 - i)/k_h = (2D_h/\omega)^{1/2} \qquad (2.3.47)$$

and called the thermal penetration length. The exponential of $-z/\delta_h$ describes a very rapid attenuation of thermal waves. For example, in argon at 300 K and 0.1 MPa $\delta_h = 82 \ \mu m$ at 1 kHz (in water under the same conditions $\delta_h = 7 \ \mu m$). The ratio δ_h/λ remains small compared with unity at frequencies that are small compared with the inverse of the thermal relaxation time τ_h (i.e. for a gas, at frequencies small compared with the molecular collision frequency).

For a shear wave, $\nabla \cdot v_r$ must vanish by definition and v_r must be a vector eigenfunction of ∇^2 with scalar eigenvalue $-k_s^2$. One such solution with 'propagation' in the positive z direction is

$$v_r = (i + j)C \exp[i\omega t - (1 + i)z/\delta_s] \qquad (2.3.48)$$

where i and j are orthogonal unit base vectors perpendicular to the z axis and δ_s is the shear penetration length given in terms of the propagation constant $k_s = (\omega/u_0)\Gamma_s$ by

$$\delta_s = (1 - i)/k_s = (2D_s/\omega)^{1/2}. \qquad (2.3.49)$$

Here, the motion is transverse to the direction of 'propagation' and again there is a very rapid attenuation.

2.4   Boundary Conditions

We have now established the fundamental equations that describe the propagation of sound in an infinite fluid medium. However, when the fluid is confined to a particular region by a surface the solutions of the wave equation must also satisfy certain boundary conditions at that surface. We begin our discussion with a general treatment of the boundary conditions that exist at the interface between two ideal media. This treatment is at a level of approximation consistent with that of §2.2 and is relevant to many problems of reflection and refraction of sound waves. We shall then consider the more general boundary conditions that must be considered for reflection of sound at the interface between a real fluid and a solid wall. In gases, the effects of thermal conduction and viscosity at such an interface can be important and it is towards that problem that the more detailed treatment is aimed.

2.4.1 Reflection and refraction in ideal media

We shall restrict our attention to the situation, illustrated in figure 2.2, of plane waves incident on a plane boundary. Let the boundary between medium I ($z \leqslant 0$) and medium II ($z > 0$) lie in the xy plane at $z = 0$. Further, let us suppose that the incident waves originate in medium I and that the motion is orthogonal to the y axis. In addition to this source wave there will be a reflected wave in medium I and a transmitted wave in medium II. There are two boundary conditions that must be satisfied. First, since there can be no net force acting on the massless boundary, the pressure must be equal on each side of the interface. Second, since the two media are assumed to remain in contact, the velocities normal to the boundary must be equal. Since we are neglecting viscosity here, there is no requirement for the tangential velocity to be a continuous function of position through the interface.

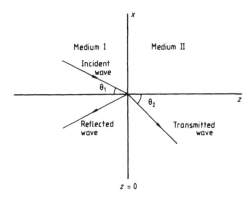

Figure 2.2 Coordinate system for plane-wave reflection and transmission at a plane surface.

We first consider medium I which is assumed to be a fluid. The acoustic pressure in the composite wave field due to both source and reflected waves is assumed to be of the form

$$p_1 = f_1(x)f_2(z)\exp(\mathrm{i}\omega t) \qquad (2.4.1)$$

in which the product $f_1 f_2$ must be a solution of the Helmholtz equation. If we set

$$\nabla^2 f_1 = -k_t^2 f_1 \qquad (2.4.2)$$

then

$$\nabla^2 f_2 = -(k_1^2 - k_t^2)f_2 \qquad (2.4.3)$$

where $k_t \leqslant k_1$ is the tangential propagation constant. Since we are neglecting dissipation, the propagation constant k_1 is simply ω/u_1, where we have used the subscript 1 to indicate medium I. The appropriate solution for f_2 is given by

$$f_2(z) = A\{\exp[-\mathrm{i}(k_1^2 - k_t^2)^{1/2}z] + \chi_{\mathrm{Re}} \exp[\mathrm{i}(k_1^2 - k_t^2)^{1/2}z]\} \quad (2.4.4)$$

in which the first term arises from the incident wave and the second from the reflected wave. In equation (2.4.4), A is an amplitude and χ_{Re} is the reflection coefficient. In view of equation (2.2.11), the z (normal) component of the velocity in medium I is given by

$$v_{1,z} = (\mathrm{i}/\omega\rho_1)(\mathrm{d}f_2/\mathrm{d}z)f_1 \exp(\mathrm{i}\omega t) \quad (2.4.5)$$

where ρ_1 is the density of medium I.

In medium II, which we first assume to be a fluid,[7] the wave transmitted through the interface is also described by an equation of the form of (2.4.1). In order to satisfy the boundary conditions at every point on the interface the dependence of the acoustic pressure p_2 on x must be identical with that of p_1. Thus we have

$$p_2 = f_1(x)f_3(z)\exp(\mathrm{i}\omega t) \quad (2.4.6)$$

and

$$\nabla^2 f_3 = -(k_2^2 - k_t^2)f_3 \quad (2.4.7)$$

where k_2 is the propagation constant in medium II. The appropriate solution here for the function f_3 corresponds to just one wave:

$$f_3 = A\chi_{\mathrm{Tr}} \exp[-\mathrm{i}(k_2^2 - k_t^2)^{1/2}z]. \quad (2.4.8)$$

Here χ_{Tr} is called the transmission coefficient. To determine the reflection and transmission coefficients we apply the two boundary conditions. The first gives

$$1 + \chi_{\mathrm{Re}} = \chi_{\mathrm{Tr}} \quad (2.4.9)$$

and the second

$$\rho_2(k_1^2 - k_t^2)^{1/2}(1 - \chi_{\mathrm{Re}}) = \rho_1(k_2^2 - k_t^2)^{1/2}\chi_{\mathrm{Tr}}. \quad (2.4.10)$$

If we now recognize that $(k_1^2 - k_t^2)^{1/2} = (\omega/u_1)\cos\theta_1$, where θ_1 is the angle of incidence relative to the normal in medium I, and that a similar relation holds in medium II, then we can simplify equation (2.4.10) before eliminating χ_{Tr} with (2.4.9) to obtain

$$\chi_{\mathrm{Re}} = \frac{r_2 \cos\theta_1 - r_1 \cos\theta_2}{r_2 \cos\theta_1 + r_1 \cos\theta_2}. \quad (2.4.11)$$

In equation (2.4.11), we have denoted the characteristic impedance ρu of a medium by a subscripted r. This expression has many interesting properties analogous with reflection and refraction of light at the interface between two media with different refractive indices. Here it is the acoustic resistances and speeds of sound that control the problem. For example, the direction θ_2 of the transmitted wave is given by

$$\cos \theta_2 = [1 - (u_2/u_1)^2 \sin^2 \theta_1]^{1/2} \qquad (2.4.12)$$

which has a real solution only if $\sin \theta_1 < u_1/u_2$. If $u_1 < u_2$ and the angle of incidence exceeds the critical angle θ_c, where $\sin \theta_c = u_1/u_2$, then there is total reflection of the incident wave. There are two limiting cases that are of special interest here; these are the extremes in which r_1/r_2 approaches either zero or infinity. In the first case, of which an example might be the reflection of sound travelling in a dilute gas from a mercury surface, χ_{Re} approaches unity; there is total reflection without change in phase and the fluid velocity vanishes at the interface. The boundary is then said to be rigid. In the second case, of which an example might be the reflection of sound waves travelling in mercury from an interface with a dilute gas, χ_{Re} approaches -1; there is again total reflection but now accompanied by a phase shift of π. In this case the acoustic pressure vanishes at the interface which is then said to have pressure release.

For the case of a boundary with a solid medium, the complete problem is complicated by the possibility of elastic shear waves propagating in the solid. However, we have here little interest in the nature of the transmitted wave and therefore simply assign to the boundary an acoustic impedance Z_a; this impedance is defined by $Z_a = p_a/v_z$. In making this assignment we assume that the motion of any element of the surface is determined solely by the acoustic pressure acting there. Such a boundary is said to be of local reaction. Actually, it will be rather more convenient to work with the dimensionless specific acoustic admittance $y_0 = \rho u/Z_a$ of the interface where ρu is the acoustic impedance of the adjacent fluid. In general y_0 will be a complex quantity with components $\xi + i\sigma$; we call ξ the specific acoustic conductance and σ the specific acoustic susceptance. The admittance is typically frequency dependent and may also depend upon the angle of incidence.

Having defined the boundary admittance it is a simple matter to show that the reflection coefficient is given by

$$\chi_{Re} = \frac{\cos \theta_1 - y_0}{\cos \theta_1 + y_0}. \qquad (2.4.13)$$

2.4.2 Thermal and viscous boundary layers

We now turn to a more detailed description of the boundary conditions for plane-wave reflection from a solid wall. We again consider a fluid medium, confined to the half space $z \leqslant 0$ by an infinite plane wall of local reaction, in which the acoustic pressure is given by equation (2.4.1). It will be sufficient to neglect the thermal and viscous terms in the propagational wave, so that $k = \omega/u$ and the other acoustic variables are given by

$$T_p = [(\gamma - 1)/\gamma\beta] f_1 f_2 \exp(i\omega t) \tag{2.4.14}$$

$$v_{p,z} = (i/\omega\rho)(df_2/dz)f_1 \exp(i\omega t) \tag{2.4.15}$$

$$v_{p,x} = (i/\omega\rho)(df_2/dz)f_2 \exp(i\omega t). \tag{2.4.16}$$

There are four boundary conditions that must be satisfied at the surface: (i) the ratio $\rho u v_z/p_a$ must equal the specific acoustic admittance y_0 of the boundary surface; (ii) the temperatures of the fluid and the wall must be equal at the surface; (iii) there must be continuity of heat flow across the interface; and (iv) the tangential component v_x of the fluid velocity must vanish at the surface. The propagational mode cannot, on its own, satisfy all four of these conditions. The second and third conditions demand that thermal waves be generated in both the fluid and the wall. It turns out that, for the cases in which thermal effects are important, the amplitude of the wave that penetrates the wall is very much less than that in the fluid so that we may usually ignore the former and simply require the latter to cancel the temperature fluctuations associated with the propagational mode. The fourth condition demands that a shear wave in the fluid be generated at the surface so as to cancel the tangential motion associated with the incident propagational wave. Because both thermal and shear waves decay very rapidly in the fluid, the regions of their importance are restricted to layers near to the boundary which are called the thermal and shear boundary layers and penetrate the gas only to depths of order δ_h and δ_s. We have assumed here that the ratios of the viscous and thermal penetration lengths to the mean free path are still large so that the boundary conditions above are those that apply to a continuum fluid. Later we shall include the modifications that arise when that condition is not met. In the bulk of the fluid only the propagational mode is important. However, the support of the thermal and shear modes in the boundary layers, with energy surrendered by the propagational wave, can have an important influence on reflection and absorption of sound at the interface between a gas and a solid.

The acoustic temperature in the fluid will be the sum of the contributions of the propagational and thermal modes. In order to satisfy the boundary conditions, the thermal wave must be described by a function that has the same dependence on x as does T_p. Thus we must have a thermal wave of the form

$$T_h = f_1(x)f_3(z)\exp(i\omega t) \qquad z \leqslant 0 \tag{2.4.17}$$

where $f_3(z)$ is a function to be determined. Since T_h is an eigenfunction of ∇^2 with eigenvalue $-k_h^2$ and f_1 is also an eigenfunction of ∇^2 but with eigenvalue $-k_t^2$ it follows that

$$\nabla^2 f_3 = -(k_h^2 - k_t^2)f_3. \qquad (2.4.18)$$

The appropriate solution is $f_3(z) = B \exp(ik_3 z)$ with $k_3 = (k_h^2 - k_t^2)^{1/2}$. However, k_t^2 is of order $(2\pi/\lambda)^2$ while k_h^2 is of order $1/\delta_h^2$; consequently the tangential propagation constant is negligible in comparison with $k_h = (1 - i)/\delta_h$ at ordinary frequencies and an excellent approximate form of the thermal wave in the fluid is

$$T_h = f_1(x)B \exp[i\omega t + (1 + i)z/\delta_h] \qquad z \leqslant 0. \qquad (2.4.19)$$

This solution, which has the property of decaying rapidly in the fluid away from the wall, will be valid when the wavelength $\lambda \gg \delta_h$; that condition is readily met. A thermal wave will also penetrate the wall material. This will be described by a function of the same form as that of equation (2.4.17) but with a dependence of z such that the thermal wave decays with increasing z. Thus we have

$$T_h = f_1(x)C \exp[i\omega t - (1 + i)z/\delta_w] \qquad z > 0 \qquad (2.4.20)$$

where δ_w is the thermal penetration length in the wall material and we have neglected k_t in comparison with $1/\delta_w$. The condition that the temperatures of the wall and the gas be equal at the surface provides a relation between the amplitudes of the two thermal waves:

$$B + [(\gamma - 1)/\gamma\beta] f_2(0) = C. \qquad (2.4.21)$$

A second relation between B and C is provided by the condition requiring continuity of heat flow. While it is possible to treat this condition exactly, the heat flows associated both with the propagational mode and with the tangential components of the thermal waves are negligible compared with the normal heat flow in the two thermal waves. This is a consequence of the ratios λ/δ_h and λ/δ_w being large compared with unity. It follows that an excellent approximation is given by equating at $z = 0$ the normal components of the heat fluxes associated just with the two thermal waves. Applying Fourier's law, and denoting the thermal conductivity of the wall material by $\varkappa_w$, this condition requires that

$$C = -(\varkappa/\varkappa_w)(\delta_w/\delta_h)B. \qquad (2.4.22)$$

Combining equations (2.4.21) and (2.4.22) we obtain the amplitude of the thermal wave in the gas:

$$B = -[(\gamma - 1)/\gamma\beta] f_2(0)[1 + (\varkappa/\varkappa_w)(\delta_w/\delta_h)]^{-1}. \qquad (2.4.23)$$

Usually, the ratio δ_w/δ_h is of order unity but for a typical fluid and a metal wall the ratio $\varkappa/\varkappa_w \ll 1$; in this case $C \ll B$ and the two conditions applied to the acoustic temperature reduce to a single condition requiring the total acoustic temperature, like the tangential fluid velocity, to vanish at the wall. The acoustic cycle, which is very nearly isentropic in the bulk of the fluid, becomes very nearly isothermal at the wall. In this approximation the amplitudes of the thermal waves are given simply by

$$B = - [(\gamma - 1)/\gamma\beta] f_2(0) \quad \text{and} \quad C = 0. \tag{2.4.24}$$

The normal derivative of the temperature field in the boundary layer is very large. Consequently, the thermal mode makes a contribution to the normal component of the fluid velocity that cannot be neglected at the surface. In view of equation (2.3.39), and neglecting thermal penetration of the wall, this contribution is given by

$$v_{h,z} = - (1 + i)(\omega\delta_h/2\rho u^2)(\gamma - 1) f_1 f_2(0) \exp[i\omega t + (1 + i)z/\delta_h] . \tag{2.4.25}$$

The contributions of the thermal mode to the acoustic pressure and to the tangential fluid velocity may be neglected to leading order in δ_h.

In order to satisfy the boundary condition pertaining to the tangential fluid velocity, we must include a shear wave near to the wall with fluid velocity v_s. Each component of v_s must be an eigenfunction of ∇^2 with eigenvalue $- k_s^2$. Clearly, we require the component of this velocity that is perpendicular to both the x and z axes to vanish. We also require the x component $v_{s,x}$ to be the negative of $v_{p,x}$ on the surface $z = 0$ so that this term is given by

$$v_{s,x} = - (i/\omega\rho)(df_1/dx) f_2(0) \exp[i\omega t + (1 + i)z/\delta_s] \tag{2.4.26}$$

where we have made the approximation $(k_s^2 - k_t^2)^{1/2} \approx k_s = (1 - i)/\delta_s$. Since the shear flow is divergence free, the x and z components of v_s must satisfy $(\partial v_{s,z}/\partial z) = - (\partial v_{s,x}/\partial x)$ which, when integrated and combined with equation (2.4.2), gives

$$v_{s,z} = - (1 + i)(k_t/k)^2(\omega\delta_s/2\rho u^2) f_1 f_2(0) \exp[i\omega t + (1 + i)z/\delta_s] . \tag{2.4.27}$$

We see that, like the thermal wave, the shear wave decays rapidly in the gas away from the wall.

In our present approximation, only the propagational wave contributes to the acoustic pressure so that $p_a = f_1 f_2 \exp(i\omega t)$. However, the total z component of the fluid velocity is given by the combination of equations (2.4.15), (2.4.25) and (2.4.27) and contains contributions from all three modes. At the surface $z = 0$ these three equations give

$$v_z = (f_1 i/\omega\rho)\{(df_2/dz)_{z=0}$$
$$- (1 - i) f_2(0)(\omega^2/2u^2)[(k_t/k)^2\delta_s + (\gamma - 1)\delta_h]\}\exp(i\omega t). \tag{2.4.28}$$

Finally, if the boundary condition $v_z = p_a(y_0/\rho u)$ is applied then an equation is obtained linking the acoustic pressure and its normal derivative at the surface $z = 0$:

$$(1/p_a)(\partial p_a/\partial n) = - i(\omega/u)(y_0 + y_h + y_s) \qquad z = 0 \tag{2.4.29}$$

where n is the wall-pointing normal ($\partial/\partial n = \partial/\partial z$). The quantities y_h and y_s, called the specific acoustic admittances of the thermal and shear boundary layers respectively, are given by

$$y_h = (1 + i)(\gamma - 1)(\omega/2u)\delta_h$$
$$= (1 + i)(\gamma - 1)u^{-1}(\pi f D_h)^{1/2} \qquad (2.4.30)$$

and

$$y_s = (1 + i)(k_t/k)^2(\omega/2u)\delta_s$$
$$= (1 + i)(k_t/k)^2 u^{-1}(\pi f D_s)^{1/2}. \qquad (2.4.31)$$

Equation (2.4.29) is a very important result. It enables us to incorporate all of the effects of heat flow and viscous damping into a single boundary condition that is to be satisfied by the propagational mode. Notice that viscous effects on the sound wave near to the wall depend, as expected, upon the angle of incidence, but that the effects of heat flow at the wall are independent of k_t. Should thermal penetration of the wall material be important then the acoustic admittance assigned to the thermal boundary layer is simply modified by the factor $[1 + (\chi/\chi_w)(\delta_w/\delta_h)]^{-1}$.

The order of magnitude of the thermal and viscous effects at a surface may be typified by the values $\text{Re}(y_h) = 0.000\,54$ and $\text{Re}(y_s) = 0.000\,66$ for sound waves at 1 kHz travelling parallel to a rigid boundary in argon at 300 K and 0.1 MPa.

It is a simple matter to calculate the time average of the power dissipated per unit area in the boundary layers. This is equal to the effective conductance $\xi/\rho u$ of the surface multiplied by the value of $|p_a|^2$ at the surface:

$$[\text{Re}(y_h + y_s)/\rho u]\,|p_a|^2_{z=0}$$
$$= (\omega/2\rho u^2)[(k_t/k)^2\delta_s + (\gamma - 1)\delta_h]\,|p_a|^2_{z=0}. \qquad (2.4.32)$$

As an example we consider the attenuation of a nominally plane wave travelling in a tube of radius b. For such a wave the ratio k_t/k is unity to a good approximation so that the power lost per unit length of the tube is

$$\delta P = (\pi b \omega/\rho u^2)[\delta_s + (\gamma - 1)\delta_h]\,|p_a|^2. \qquad (2.4.33)$$

Now, since the RMS power P transmitted along the tube by a plane wave is equal to the cross-sectional area multiplied by the acoustic intensity, the fraction of the power dissipated per unit length in the boundary layers, $\delta P/P$, is $(\omega/ub)[\delta_s + (\gamma - 1)\delta_h]$. Thus the extra absorption due to the boundary layer losses leads to a contribution

$$\alpha_{KH} = (\omega/2ub)[\delta_s + (\gamma - 1)\delta_h] \qquad (2.4.34)$$

to the effective absorption coefficient for plane waves travelling along the tube; this is known as the Kirchhoff–Helmholtz absorption coefficient.[8] At low frequencies in narrow tubes the boundary layer losses can far exceed those due to absorption in the bulk of the fluid. For example, for argon gas at 300 K, 0.1 MPa and 1 kHz in a tube of 1 cm diameter, $\alpha_{KH} = 0.24$ m^{-1} which, while small compared with $\omega/u = 19.5$ m^{-1}, greatly exceeds $\alpha_{cl} = 2 \times 10^{-5}$ m^{-1}.

The non-zero boundary layer admittance will actually perturb both real and imaginary parts of the propagation constant and, according to the Kirchhoff–Helmholtz theory (which neglects the bulk absorption), the propagation constant is given by

$$k = (\omega/u) + (1 - i)\alpha_{KH} \qquad (2.4.35)$$

where u is the speed of sound in free space [6].[9] This shows that the speed of nominally plane waves travelling along the tube, equal to the real part of ω/k, will fall short of that in free space by a fraction $\alpha_{KH}u/\omega$.

There have been many experimental tests of the Kirchhoff–Helmholtz theory reported in the literature. Most workers are in agreement with regards to the dependence upon frequency and pressure in a gas although some have found values of the absorption coefficient up to 20 per cent greater than that predicted [7, 8] (it is likely that such discrepancies are attributable to experimental error). The comprehensive results obtained by Quinn *et al* [9] in a tube of 30 mm diameter operated at 5.6 kHz in argon at 273.16 K and pressures between 10 and 200 kPa yielded values of the absorption coefficient within about 1.5 per cent of those predicted from the theory using the available transport and equilibrium properties of the gas. More generally, the boundary layer theory has been tested for the radial modes of spherical resonators [10, 11] and for radial and longitudinal modes of cylindrical resonators [12]. In the latter case, the experiments were designed to exploit the boundary layer losses in a determination of η and $\varkappa$. The results near to 300 K and 0.1 MPa were in agreement with accepted values to better than 0.5 per cent in the gases studied.

2.4.3 Molecular slip and the temperature jump

The boundary conditions of the previous section were those pertaining to a continuum fluid. In gases at low pressures the finite length of the mean free path becomes important and hydrodynamics can no longer be applied with confidence. Under usual conditions, the shortest macroscopic lengths involved in the problem are the thermal and shear penetration lengths and so, as the pressure of a gas is reduced, the first place that deficiencies in classical hydrodynamics would be expected to show up would be here in the theory of the boundary layers. These expectations are borne out in experiment [11]. The theoretical description of heat and momentum transfer near to an interface under low-pressure conditions is less well developed than in the hydrodynamic regime. However, a treatment based on very simple kinetic theory is consistent with the experimental results. We first consider momentum transfer.

Let the gas be confined to the half space $z \leqslant 0$ by an infinite plane wall as before. We suppose that there is a tangential velocity v_t that is a linear function of z in the bulk of the gas. However, we now recognize that, because of the molecular nature of the fluid, the tangential velocity need not vanish in the gas immediately adjacent to the wall; instead the situation shown in figure 2.3 prevails in which

$$v_t(z) = v_0 - z(\mathrm{d}v_t/\mathrm{d}z) \qquad (2.4.36)$$

and v_0 is called the slip velocity. Note that the classical boundary layer theory does indeed lead to a linear variation of $v_t(z)$ close to the wall (i.e. for $|z| \ll \delta_s$) although it assumes $v_0 = 0$. In order to determine v_0, we apply simple kinetic theory to the momentum flow in the layer within a few mean free paths of the wall.

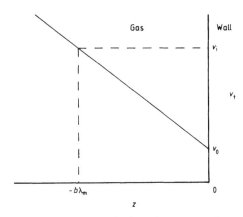

Figure 2.3 Tangential fluid velocity close to a boundary in the molecular slip regime.

According to kinetic theory, the rate of arrival of molecules per unit area at the surface is given by $\frac{1}{4}\bar{n}\langle c \rangle$ where $\bar{n}$ is the number density, $\langle c \rangle = (8kT/\pi m)^{1/2}$ is the mean molecular speed, and m is the mass of one molecule. The mean tangential velocity v_i of the incident molecules corresponds to that at a distance $b\lambda_m$ from the wall where λ_m is the mean free path and b a constant of order unity. Thus the tangential momentum current density carried to the wall by the incident molecules in $\frac{1}{4}v_i\rho\langle c \rangle$. The mean tangential velocity of the molecules emitted from the wall should lie in the range between zero and v_i depending on the extent to which they come to equilibrium with the wall before being re-emitted. Let this mean tangential velocity of the emitted molecules be $(1 - h_s)v_i$, where h_s is the momentum accommodation coefficient and takes values between zero (specular reflection) and unity (diffusive reflection). Thus the magnitude of the net momentum current density at the interface is given by

$$J_P = \tfrac{1}{4} h_s \rho \langle c \rangle v_i = \tfrac{1}{4} h_s \rho \langle c \rangle [v_0 - b\lambda_m (dv_t/dz)]. \qquad (2.4.37)$$

Away from the interface in the bulk of the gas molecules are free to cross the xy plane from both directions and those crossing in either direction carry a tangential velocity typical of that a distance $b\lambda_m$ behind the plane and arrive at a rate $\tfrac{1}{4} \bar{n} \langle c \rangle$ per unit area. Consequently, the net momentum current density there is

$$J_P = -\tfrac{1}{2} b\lambda_m \rho \langle c \rangle (dv_t/dz) \qquad (2.4.38)$$

and we can identify $\eta = \tfrac{1}{2} b\lambda_m \rho \langle c \rangle$ as the kinetic theory expression for the coefficient of shear viscosity. The fundamental boundary condition is that requiring continuity of momentum current in the gas. Since we have assumed a linear variation of tangential velocity with z we can equate the momentum flux at the wall with that a few mean free paths away; this serves to determine the slip velocity through equations (2.4.37) and (2.4.38):

$$v_0 = -b\lambda_m [(2 - h_s)/h_s] (dv_t/dz) = b\lambda_m [(2 - h_s)/h_s] (J_P/\eta) \qquad (2.4.39)$$

(note that the momentum current has the direction of the wall-pointing normal). Finally, $b\lambda_m$ may be eliminated in favour of η using equation (2.4.38) to obtain

$$v_0 = l_s (J_P/\eta) \qquad (2.4.40)$$

where l_s, called the momentum accommodation length, is given by

$$l_s = (\eta/p)(\pi RT/2M)^{1/2} [(2 - h_s)/h_s]. \qquad (2.4.41)$$

The accommodation length is of the order of the mean free path when the accommodation coefficient is near unity. This appears to be the case in practice unless the surface is exceptionally smooth and completely degassed in high vacuum prior to exposure to the gas. Unfortunately it is not possible to evaluate the accommodation coefficient *a priori*.

Using similar arguments it is possible to evaluate the 'temperature jump' that occurs at the interface between a dilute gas and a solid in the presence of a heat flux. The temperature of the gas T_e extrapolated to the wall then differs from T_w, that of the wall itself, by

$$T_e - T_w = l_h (J_h/\varkappa) \qquad (2.4.42)$$

where J_h is the magnitude of the wall-pointing heat current density and l_h is the thermal accommodation length; for a monatomic gas l_h is given by

$$l_h = (\varkappa/p)(\pi MT/2R)^{1/2} [(2 - h_h)/2h_h] \qquad (2.4.43)$$

where h_h is the thermal accommodation coefficient. The kinetic theory of heat transfer in polyatomic gases is complicated by the important role of inelastic collisions but elementary theory suggests that equation (2.4.43) should be modified so that

$$l_h = (\varkappa/p)(\pi MT/2R)^{1/2}[1 + \tfrac{1}{2}(C_{m,\text{int}}/R)]^{-1}[(2 - h_h)/2h_h] \quad (2.4.44)$$

for a polyatomic gas with internal molar heat capacity $C_{m,\text{int}}$.

For argon at 300 K and 0.1 MPa at a surface with accommodation coefficients equal to unity, $l_s = 71$ nm and $l_h = 133$ nm; both are usually small compared with δ_s and δ_h at frequencies where boundary layer effects are important. However, since the mean free path increases faster than do the penetration lengths as the pressure is reduced, a very-low-pressure regime exists where the boundary layer theory cannot be expected to hold.

Having derived these modified boundary conditions it is a simple matter to adjust the amplitudes of the thermal and shear waves in the gas so that they are obeyed. We now require $T_p + T_h = l_h(J_h/\varkappa)$ and $v_{p,x} + v_{r,x} = l_s(J_p/\eta)$ at $z = 0$ in place of the conditions requiring both the acoustic temperature and the tangential velocity in a dilute gas to vanish at the interface. Both the heat flux and the momentum flux may be evaluated neglecting the propagational mode's contributions. The result of this treatment is that the amplitude of the thermal wave is modified by the factor $[1 - (1 + i)(l_h/\delta_h)]$, correct to leading order in the small ratio l_h/δ_h. Similarly, the amplitude of the shear wave is modified by the factor $[1 - (1 + i)(l_s/\delta_s)]$ correct to leading order in l_s/δ_s. It then follows that the specific acoustic admittances assigned to the thermal and shear boundary layers, equations (2.4.30) and (2.4.31), are modified by the same factors so that

$$y_h = (\gamma - 1)(\omega/2u)[(1 + i)\delta_h - 2il_h] \quad (2.4.45)$$

$$y_s = (k_t/k)^2(\omega/2u)[(1 + i)\delta_s - 2il_s]. \quad (2.4.46)$$

These equations show that inclusion of the slip or temperature-jump term leaves the conductance of a boundary layer unchanged but reduces the susceptance by a fraction equal to twice the ratio of the accommodation length to the penetration length. Thus the power dissipated in the boundary layers remains unchanged.

2.4.4 Condensation and precondensation phenomena

In discussing the reflection of sound at the boundary between two media we have tacitly assumed that there is no transfer of material across the interface. At first glance this would appear to be a reasonable assumption for reflection at a solid surface. However, for the case of reflection from the interface between coexisting vapour and liquid phases this assumption is clearly invalid; the variation of temperature and pressure during the acoustic cycle will cause evaporation and condensation at the interface and give rise to a non-zero normal fluid speed. A similar situation arises at a solid surface coated by a film of condensed vapour. Such precondensation effects are known to enhance greatly the surface admittance as the pressure of the gas is increased towards saturation [13].

It is not difficult to evaluate the specific acoustic admittance for a sound wave impinging on the interface between coexisting vapour and liquid phases. Let us assume that the interface lies in the xy plane at $z = 0$ and that the vapour occupies the half space $z < 0$. The effects of vaporization greatly exceed those of the ordinary thermal and shear boundary layers which we shall ignore in the following. Consequently, we neglect the tangential velocity entirely and restrict our attention to the propagational and thermal waves on both sides of the interface. Four boundary conditions are required to specify a solution. We shall assume that the density of the vapour is small compared with that of the liquid so that to a good approximation the acceleration of the liquid can be neglected. We therefore require continuity of pressure, temperature and heat flux at $z = 0$; in addition, we shall assume that local thermodynamic equilibrium is maintained so that

$$p_a = (dp^\sigma/dT)T_a \qquad (2.4.47)$$

at the interface, where p^σ is the saturated vapour pressure. The various contributions to the acoustic pressure, temperature and normal fluid speed in the two phases will be given by:

$$
\begin{aligned}
p_{p,g} &= f_1(x)f_2(z)\exp(i\omega t) & p_{p,l} &= f_1(x)f_3(z)\exp(i\omega t) \\
T_{p,g} &= [(\gamma_g - 1)/\gamma_g\beta_g]\,p_{p,g} & T_{p,l} &= [(\gamma_l - 1)/\gamma_l\beta_l]\,p_{p,l} \\
v_{z,g} &= (i/\omega\rho_g)f_1(df_2/dz)\exp(i\omega t) & v_{z,l} &= (i/\omega\rho_l)f_1(df_3/dz)\exp(i\omega t) \\
T_{h,g} &= B_g\,\exp[i\omega t + (1+i)z/\delta_{h,g}] & T_{h,l} &= B_l\,\exp[i\omega t + (1+i)z/\delta_{h,l}]
\end{aligned}
\qquad (2.4.48)
$$

where subscripts g and l denote properties of the gas and liquid. Consistent with our neglect of terms of the order of the ordinary boundary layer effects, the normal fluid speeds given in equations (2.4.48) do not include contributions from the thermal modes. In addition we are assuming that $v_{z,l} \approx 0$ so that the interface is essentially static and the mass flux there is $\rho_g v_g$. The material periodically evaporating and condensing at the interface contributes a power per unit area equal to $\rho_g v_g \Delta h$, where Δh is the specific enthalpy of vaporization. As before, we neglect the heat flux associated with the propagational waves in comparison with the much larger contributions from the thermal waves. Applying Fourier's law, the boundary condition on the normal heat flux therefore requires that

$$(1 + i)[\,(\varkappa_g/\delta_g)B_g + (\varkappa_l/\delta_l)B_l\,] = (i/\omega)\Delta h f_1(x)(df_2/dz)_{z=0}. \quad (2.4.49)$$

Under the conditions where $\rho_g \ll \rho_l$, $(\varkappa_g/\delta_g)B_g$ will be much smaller than $(\varkappa_l/\delta_l)B_l$ and can be neglected. This means that the flow of heat to and from the interface is dominated by that in the liquid so we need consider no further the thermal wave in the vapour. In addition, the large value of $\beta = (\partial p/\partial T)_\rho$ for the liquid means that $T_{p,l}$ is very much smaller than $T_{h,l}$ so that the former may also be neglected. Equation (2.4.47) then provides a relation between the amplitudes of the thermal and propagational waves at the liquid surface:

$$B_1 = f_1(x)f_3(0)(dp^\sigma/dT)^{-1} \qquad (2.4.50)$$

where equality of pressure implies that $f_2(0) = f_3(0)$. Finally, combining equations (2.4.49) and (2.4.50) we obtain for the gas phase

$$-(1/f_2)(df_2/dz) = iky \qquad (2.4.51)$$

where the specific acoustic admittance y is given by

$$y = (1 + i)u_g(x_l/\delta_l\Delta h)(dp^\sigma/dT)^{-1}. \qquad (2.4.52)$$

As a numerical example, we consider the reflection of sound travelling in water vapour from the surface of the coexisting liquid at 373.15 K. In this case, $u_g \approx 480 \text{ m s}^{-1}$, $x_l = 0.60 \text{ W m}^{-1}\text{K}^{-1}$, $\Delta h = 2.26 \text{ MJ kg}^{-1}$, $D_{h,l} = 1.44 \times 10^{-5} \text{ m}^2\text{s}^{-1}$ and $(dp^\sigma/dT) = 3.6 \text{ kPa K}^{-1}$ so that $\text{Re}(y) = 0.005$ at 1 kHz. This value exceeds the specific acoustic conductance for a normally incident wave in the absence of mass transfer by a factor of about 30 (both have the same frequency dependence). At low temperatures, where $(dp^\sigma/dT) \to 0$, equation (2.4.52) diverges. However, in that regime the assumptions of local equilibrium and of a hydrodynamic description begin to fail and a description including jump coefficients is required [14]. In that description, the divergence of y at low temperatures is suppressed.

The important case of an adsorbed liquid film on a solid surface has been studied experimentally by and Mehl and Moldover [13] who also derived a theoretical description based on a model for the adsorption isotherm. Their theory reduces to equation (2.4.52) for very thick films and actually predicts values of y up to one order of magnitude greater for pressures just below saturation. Experimentally, a very large enhancement of the surface admittance was observed when the pressure of propane gas was increased towards 99.5 per cent of the saturated vapour pressure; both real and imaginary components of y reached values comparable to but not significantly larger than that expected for the bulk liquid. The results were nevertheless in qualitative agreement with the theory. The lack of close quantitative agreement was probably due to the fact that, while the model assumed a perfectly smooth surface, machined metal surfaces were studied in practice. One must conclude that precondensation effects are a potential source of systematic error in many experiments designed to measure the speed and absorption of sound. In the absence of accurate theory, the experimentalist must either avoid conditions close to saturation or fall back on the general observation that the consequences of precondensation on measurements of the sound speed appear to drop off quite rapidly with increasing frequency. Thus high-frequency measurements may be expected to offer more reliable results along the phase boundaries.

References

[1] Landau L D and Lifshitz E M 1959 *Course of Theoretical Physics Volume 6. Fluid Mechanics* English edn (Oxford: Pergamon) pp 270–2

[2] Landau L D and Lifshitz E M 1959 *Course on Theoretical Physics Volume 6. Fluid Mechanics* English edn (Oxford: Pergamon) pp 47–8

[3] Greenspan M 1956 *J. Acoust. Soc. Am.* **28** 644

[4] Pinkerton J M M 1947 *Nature* **160** 128

[5] Helmholtz H 1863 *Verh. naturhist. med. Ver. Heidelberg* **3** 16

[6] Kirchhoff G 1868 *Ann. Phys. Chem. (Fifth Ser.)* **134** 177; English transl. Lindsay R B *Benchmark Papers in Acoustics: Physical Acoustics* (Stroudsburg, PA: Dowden, Hutchinson & Ross) pp 7–19

[7] Smith D H and Harlow R G 1963 *Br. J. Appl. Phys.* **14** 102

[8] Harlow R G and Kitching R 1964 *J. Acoust. Soc. Am.* **36** 1100

[9] Quinn T J, Colclough A R and Chandler T R D 1976 *Phil. Trans. R. Soc.* A **283** 367

[10] Moldover M R, Mehl J B and Greenspan M 1986 *J. Acoust. Soc. Am.* **79** 253

[11] Ewing M B, McGlashan M L and Trusler J P M 1986 *Metrologia* **22** 93

[12] Carey C, Bradshaw J, Lin E and Carnevale E H 1974 *Experimental Determination of Gas Properties at High Temperatures and/or High Pressures* (Arnold Engineering Development Center, Arnold Air Force Station, TN 37389, USA) *Report no.* AEDC-TR-74-33, also available as NTIS AD-779772

[13] Mehl J B and Moldover M R 1982 *J. Chem. Phys.* **77** 455

[14] Robnik M, Kuscer I and Lang H 1979 *Int. J. Heat Mass Transfer* **22** 461

General reference

Morse P M and Ingard K U 1968 *Theoretical Acoustics* (New York: McGraw-Hill) ch 6

Notes

1. In the Cartesian coordinate system the vector operator ∇ is written $[i(\partial/\partial x) + j(\partial/\partial y) + k(\partial/\partial z)]$, where i, j, k are the x, y, z-pointing unit base vectors.

2. For the x component of v we have

$$(dv_x/dt)\,dt = v_x(x + u_x\,dt, t + dt) - v_x(x, t)$$
$$= v_x(x, t) + (\partial v_x/\partial x)u_x\,dt + (\partial v_x/\partial t)\,dt - v_x(x, t)$$

from which it follows that

$$(\mathrm{d}v_x/\mathrm{d}t) = (\partial v_x/\partial t) + (\partial v_x/\partial x)u_x.$$

Generalization to flow in three dimensions gives equation (2.2.5).

3. Provided that the initial motion is irrotational.
4. Also known as the volume viscosity.
5. Here we use the result $\nabla^2 v_j = [\nabla(\nabla \cdot v)]_j - [\nabla \times (\nabla \times v)]_j$ where $\nabla \cdot v = 0$ for the rotational flow.
6. We include terms of the second order in $\omega \tau_h$ because this quantity appears in the dominator of equation (2.3.31).
7. In a solid medium both elastic shear waves and longitudinal compression waves are permitted; only the latter are considered here.
8. The effect of viscosity on the propagation of sound in narrow tubes was first deduced by Helmholtz [5]; the additional effects of heat conduction were included later by Kirchhoff [6].
9. For a proof of this result, see §3.6.2.

3

Cavities

3.1 Introduction

When sound is generated continuously within a closed cavity, a steady state is attained in which the wave motion is that of a standing wave. If the frequency of the source happens to coincide with a natural frequency of the system then resonance will occur. When the enclosure is of simple geometry, and the properties of the walls are known, solutions of the wave equation can be found that satisfy the boundary conditions. These provide expressions which relate the resonance frequencies and linewidths to the speed and absorption of sound in the medium, and thereby form a basis for the measurement of those quantities.

In this chapter, we will derive expressions to describe both free and forced oscillations of a fluid medium contained within a cavity; a number of simple geometries will be considered explicitly. An exact solution, even for simple geometries, would be very cumbersome and it is much more profitable to pursue an alternative route by which the wave equations are solved first for a cavity of idealized properties. Once solutions of that kind are available it turns out to be relatively straightforward to include many of the complexities of real systems as perturbations to the simplified model.

3.2 The Normal Modes of an Acoustic Cavity

Since the time dependence of a simple-harmonic standing wave is spatially uniform, the velocity potential Ψ for the region $\mathfrak{R}$ of the cavity may be separated into the product

$$\Psi(r, t) = A\Phi(r)\exp(i\omega t) \tag{3.2.1}$$

where $\omega/2\pi$ is the frequency, $\Phi(r)$ is a dimensionless wavefunction that gives the spatial variation of the wave field, and A is a constant that determines its overall amplitude. The wave equation,

$$[\nabla^2 + (k/\omega)^2(\partial^2/\partial t^2)]\Psi(r, t) = 0 \qquad (3.2.2)$$

will be satisfied if

$$\nabla^2\Phi(r) = -k^2\Phi(r) \qquad (3.2.3)$$

where

$$k = (\omega/u) - i\alpha \qquad (3.2.4)$$

is the propagation constant in the medium that fills the enclosure. The solutions of equation (3.2.3) that are allowed within the closed region $\mathfrak{R}$ are the eigenfunctions of ∇^2 that satisfy the boundary conditions at the surface $\mathfrak{S}$ of the enclosure; the corresponding eigenvalues are the allowed values of $-k^2$.

We shall seek an infinite set of solutions that are mutually orthogonal, finite and continuous within $\mathfrak{R}$; these define the normal modes of the cavity and may be used to construct any possible solution of the wave equation that is physically significant. Since the problem is three dimensional, a set of three indices is generally required to distinguish the members of this set of solutions. The eigenfunctions and their corresponding eigenvalues will therefore be denoted by $\Phi_N(r, \omega)$ and $-K_N^2(\omega)$ respectively, where N represents the trio of indices n_1, n_2, n_3. The frequency dependence of the normal-mode solutions implied by this notation may enter through the boundary conditions.

Each of the normal modes will obey the orthogonality condition

$$\iiint\limits_{\mathfrak{R}} \Phi_N(r, \omega)\Phi_M^*(r, \omega) \, dV = V\Lambda_N(\omega)\delta(N - M) \qquad (3.2.5)$$

where V is the volume of the cavity, $\Lambda_N(\omega)$ is a normalization constant, and $\delta(N - M)$ is taken to mean the product $\delta(n_1 - m_1)\delta(n_2 - m_2)\delta(n_3 - m_3)$ of three Dirac delta functions.[1]

It will be assumed that the surface $\mathfrak{S}$ of $\mathfrak{R}$ is one of local reaction so that the boundary condition is

$$v_n(r_S) = p_a(r_S)Y(r_S, \omega) \qquad (3.2.6)$$

where $v_n(r_S)$ is the outward-pointing normal component of the fluid velocity at position r_S on $\mathfrak{S}$, and $Y(r_S, \omega) = y(r_S, \omega)/\rho u$ is the effective acoustic admittance of the boundary at angular frequency ω. Thus equation (3.2.6) requires that the wavefunctions satisfy

$$(\partial/\partial n)\Phi_N(r, \omega)|_{r = r_S} = -i(\omega/u)\Phi(r_S, \omega)y(r_S, \omega). \qquad (3.2.7)$$

The effect of the boundary condition is, of course, to restrict the propagation constant k to the discrete set of values $K_N(\omega)$. Since k is also required to satisfy equation (3.2.4), the natural frequencies of the system are complex quantities given by

$$F_N = (f_N + i g_N) = (u/2\pi)(K_N + i\alpha). \tag{3.2.8}$$

Free oscillations may occur only at these discrete natural frequencies of the system—each mode oscillating in proportion to $\exp(2\pi i F_N t)$. Thus in the time domain, $1/2\pi g_N$ is the time constant with which free oscillations of the Nth mode decay. When the boundary conditions are frequency dependent, K_N should be evaluated for each mode at the corresponding angular frequency $2\pi f_N$.

3.3 Forced Oscillations

The normal modes define only the form of the free oscillations that are allowed within the cavity and, as yet, nothing about forced oscillations. However, all of the acoustic properties of the cavity, including both the transient and the steady-state response to a source of sound, can be derived from the information contained in the normal-mode solutions. It is convenient to consider first the response of the cavity to a continuous simple-harmonic source of infinitesimal size. The response to a source of finite size may then be obtained by summing the effects of infinitesimal sources, and the transient response may be obtained using the Fourier transform.

The velocity potential for the cavity driven in the steady state by an infinitesimal source of strength S_ω placed at r_0 will be denoted by

$$\Psi_\omega(r \,|\, r_0, t) = S_\omega G_\omega(r \,|\, r_0) \exp(i\omega t) \tag{3.3.1}$$

where the function $G_\omega(r \,|\, r_0)$, defining the spatial distribution of the driven wave field, is to be determined. $\Psi_\omega(r \,|\, r_0, t)$ differs from the velocity potential of the undriven cavity because it is discontinuous at the source point. Thus $G_\omega(r \,|\, r_0)$ is not a solution of the homogeneous wave equation; it can be shown (and will be in Chapter 5, §5.2.3) that G_ω is in fact a solution of the inhomogeneous equation

$$(\nabla^2 + k^2) G_\omega(r \,|\, r_0) = -\delta(r - r_0). \tag{3.3.2}$$

We require that G_ω also satisfy the boundary condition analogous to equation (3.2.7):

$$(\partial/\partial n) G_\omega(r \,|\, r_0)\,|_{r = r_S} = -i(\omega/u) G_\omega(r \,|\, r_0) y(r_S, \omega). \tag{3.3.3}$$

Whatever the form of G_ω, it may be expanded as an infinite series of orthogonal functions. The natural choice is the set of normal modes of the cavity; we therefore write

$$G_\omega(r \,|\, r_0) = \sum_N C_N \Phi_N(r, \omega). \tag{3.3.4}$$

In order to evaluate the coefficients of the series we substitute it in equation (3.3.2) obtaining

$$\sum_N C_N \Phi_N(r, \omega)[K_N^2(\omega) - k^2] = \delta(r - r_0) \tag{3.3.5}$$

which, on multiplication by $\Phi_M^*(r, \omega)\,dV$ and integration over $\mathfrak{R}$, yields the following expression for a general coefficient of the series:

$$C_M = \Phi_M^*(r_0, \omega)/V\Lambda_M[K_M^2(\omega) - k^2]. \tag{3.3.6}$$

$G_\omega(r \,|\, r_0)$ is the Green function for the driven cavity. Its expansion in terms of the normal modes

$$G_\omega(r \,|\, r_0) = \sum_N \frac{\Phi_N(r, \omega)\Phi_N^*(r_0, \omega)}{V\Lambda_N[K_N^2(\omega) - k^2]} \tag{3.3.7}$$

has ensured that the boundary conditions are automatically satisfied. An important property of this function is that it is symmetric with respect to exchange of the source point r_0 and a measurement point r. In view of the appearance of the complex conjugate in the numerator of equation (3.3.7), this symmetry may not be immediately apparent. However, when there are complex wavefunctions Φ_N and Φ_N^* will be a degenerate pair, each with eigenvalue $-K_N^2$, and both will appear in the series thus preserving its symmetry with respect to r and r_0.

The general formula for the spatial distribution of the acoustic pressure $p_\omega(r \,|\, r_0)$ in the driven cavity is obtained by operating on Ψ_ω with $\rho(\partial/\partial t)$ and dropping the time-dependent factor; the result of this operation is

$$p_\omega(r \,|\, r_0) = i\omega\rho S_\omega \sum_N \frac{\Phi_N(r, \omega)\Phi_N^*(r_0, \omega)}{V\Lambda_N[K_N^2(\omega) - k^2]} \tag{3.3.8}$$

where $k = (\omega/u) - i\alpha$ and $\omega = 2\pi f$ is the (real) angular frequency of the source. The acoustic pressure is a complex quantity because the response of the driven cavity is not necessarily in phase with the source.

The denominator of the Nth term in the series for G_ω is a minimum for $\mathrm{Re}(k^2) = \mathrm{Re}(K_N^2)$ so that the mode is resonant when the frequency of the source is near to f_N, the real part of the complex natural frequency F_N defined by equation (3.2.8). When the modes are well resolved, a single component (or a group of degenerate components) predominates near resonance. The positive imaginary component of K_N, arising from dissipation at the boundary, and the negative imaginary component of k, arising from dissipation in the bulk of the fluid, combine to prevent the denominator of equation (3.3.8) from ever becoming zero and are responsible for finite resonance linewidths.

One can show that the mode density (mean number of natural frequencies per unit bandwidth) increases with the frequency, at first as an irregular function but ultimately smoothly as order f^2 [1]. Thus at low frequencies, the response of the driven cavity is punctuated by more or less well resolved resonances, while at high frequencies the resonances tend to merge and eventually overlap into a continuum response. In the former case, a single mode, or a group of degenerate modes, can be studied in near isolation and an approximation to equation (3.3.8) is appropriate in which the summation is extended over a much smaller number of terms and the 'background' contribution of the other modes is expanded in a Taylor series about some frequency f_0 near to the resonances of interest:

$$p_\omega = \sum_{N=N_1}^{N_2} \left(\frac{A_N}{(F_N/f)^2 - 1} \right) + B + C(f - f_0) + \cdots. \qquad (3.3.9)$$

Here we have also made use of the approximation $(u/2\pi)^2(K_N^2 - k^2) \approx F_N^2 - f^2$ near resonance. The summation may include just a single mode, when well resolved, or a small group of degenerate or nearly degenerate modes. The coefficients $A_N, B, C, \cdots$ are taken as complex constants for a fixed location of source and detector and over a limited frequency range.[2] The complex natural frequencies and the other constants in equation (3.3.9) may be determined from measurements of the amplitude and phase of the acoustic pressure over a small frequency band near to the resonance. A condition for precise measurements is that the quality factor $Q_N = f_N/2g_N$ of the mode or modes in question be high. This being the case, a simpler approximation for the denominator of the Nth term in the series for p_ω is possible because

$$(u/2\pi)^2(K_N^2 - k^2) \approx 2if_N[g_N + i(f - f_N)] \qquad (3.3.10)$$

correct to leading order in Q_N^{-1}. Thus for the case of a sharp singlet without background we have a Lorentzian lineshape:

$$p_\omega = \frac{ia_N}{g_N + i(f - f_N)}. \qquad (3.3.11)$$

The response of the cavity then reaches a maximum value of a_N/g_N when the source frequency is coincident with the resonance frequency. The amplitude is reduced to $1/\sqrt{2}$ of its maximum at $f = f_N \pm g_N$ and, since these are the half-power points, g_N is referred to as the half-width of the resonance and $2g_N$ is the usual measure of the linewidth. The amplitude and phase of p_ω are illustrated for a Lorentzian resonance in figure 3.1. The difference between the Lorentzian function and the resonance function of equation (3.3.9) is usually negligible but background terms are often important. These background terms can shift the apparent resonance frequency (the frequency at which the amplitude is a maximum) and their neglect could be a source of systematic error in acoustic measurements.

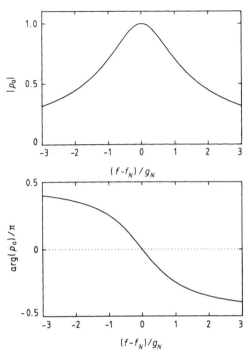

Figure 3.1 Amplitude and phase for a Lorentzian resonance. The phase is plotted relative to an arbitrary zero at the resonance frequency.

3.4 Boundary Conditions

In the previous sections, the response of a cavity to a steady simple-harmonic source of sound has been derived in general terms. In order to exploit these results we must find the eigenfunctions of the Helmholtz operator that satisfy the boundary conditions characterizing the particular cavity in question. However, if the geometry of the cavity is not particularly simple, or if the surface admittance is irregular, then eigenfunctions that obey the boundary conditions cannot be found in closed form. Usually, a simple geometry will be chosen and steps taken to secure a small and uniform surface admittance. Nevertheless, infinitely fine tolerances cannot be achieved, transducers for the generation and detection of sound usually introduce some non-uniformity, and the effective surface admittance changes with the properties of the fluid that fills the cavity. Consequently, the most efficient mathematical approach to the problem is to find first solutions for cavities of idealized properties ('perfect' geometry and zero surface admittance) and then evaluate the effects of imperfect geometry and non-zero and non-uniform surface admittance as perturbations.

We must first choose a coordinate system appropriate to the boundary shape of the idealized cavity. In this context an 'appropriate' coordinate system is one in which the wavefunctions that satisfy the boundary conditions are separable into a product of three terms each of which is a function of just one coordinate. For these specialized cases the wavefunctions have nodal surfaces (surfaces on which $\nabla\phi = 0$) that coincide with the families of coordinate surfaces. There are just 11 coordinate systems for which the wave equation is separable in three dimensions; thus even the simplified approach is not suitable in every case.[3] However, the coordinate systems that are separable correspond to the simple geometries that we would in any case choose, so the general approach is efficient and appropriate to the problem.

We now turn to the development of a general perturbation treatment that can subsequently be applied to any cavity for which solutions can be found in the idealized case. For this idealized case in which perfect geometry and zero admittance are both assumed we denote the eigenfunctions of the Helmholtz operator by $\phi_N(r)$, the corresponding eigenvalues by $-k_N^2$ and the normalization constants by Λ_N^0. Thus:

$$(\nabla^2 + k^2)\phi(r) = 0 \qquad \text{within } \mathcal{R}_0 \qquad (3.4.1)$$

$$(\partial/\partial n)\phi(r)|_{r=r_s} = 0 \qquad \text{on } \mathcal{S}_0 \qquad (3.4.2)$$

where $\mathcal{R}_0$ and $\mathcal{S}_0$ are the region and surface of the ideal cavity. The boundary conditions here are called homogeneous Neumann conditions and the eigenvalues are purely real quantities independent of frequency. The wavefunctions are defined as zero outside of the closed region $\mathcal{R}_0$. The Green function $G(r \,|\, r_0)$ that represents for this system the spatial distribution of the wave field from a point source at r_0 will be a solution of the corresponding inhomogeneous equation, subject to the same homogeneous boundary conditions:

$$(\nabla^2 + k^2)G(r \,|\, r_0) = -\delta(r - r_0) \qquad \text{within } \mathcal{R}_0 \qquad (3.4.3)$$

$$(\partial/\partial n)G(r \,|\, r_0)|_{r=r_s} = 0 \qquad \text{on } \mathcal{S}_0. \qquad (3.4.4)$$

Again we may develop $G(r \,|\, r_0)$ as an expansion in terms of the solutions of the homogeneous wave equation with the result

$$G(r \,|\, r_0) = \sum_N \frac{\phi_N(r)\phi_N^*(r_0)}{V\Lambda_N^0(k_N^2 - k^2)}. \qquad (3.4.5)$$

We now seek to find solutions of the homogeneous wave equation

$$(\nabla^2 + k^2)\Phi(r) = 0 \qquad \text{within } \mathcal{R} \qquad (3.4.6)$$

subject to the *in*homogeneous boundary conditions that prevail on the surface S of the real cavity

$$(\partial/\partial n)\Phi(r)|_{r=r_S} = -i(\omega/u)\Phi(r)y(r) \qquad \text{on } S. \qquad (3.4.7)$$

However, we wish to express the required solutions in terms of those for the idealized or unperturbed cavity with surface S_0 (not necessarily coincident with S) at which homogeneous Neumann conditions prevail; the Green function of equation (3.4.5) provides the tool that makes this possible. There is, however, one restriction that must be borne in mind: the unperturbed wavefunctions ϕ_N, which form the basis set in terms of which we seek to expand Φ_N, are (by definition) zero outside the surface S_0. Consequently, the surface S must be chosen such that it is entirely enclosed by S_0.

To proceed, we write equation (3.4.4) for k equal to an eigenvalue K_N of the perturbed cavity, denoting the Green function by $G_N(r|r_0)$, and multiply by a particular solution $\psi_N(r)$ of the homogeneous equation that corresponds to the chosen eigenvalue:

$$\psi_N(r)\nabla^2 G_N(r|r_0) + \psi_N(r)K_N^2 G_N(r|r_0) = -\delta(r-r_0)\psi_N(r). \qquad (3.4.8)$$

Next, we write equation (3.4.6) for the same particular solution, and multiply this equation by $G_N(r|r_0)$:

$$G_N(r|r_0)\nabla^2\psi_N(r) + G_N(r|r_0)K_N^2\psi(r) = 0. \qquad (3.4.9)$$

We now combine these two equations, the terms in K_N being eliminated, and interchange r and r_0 (making use of the symmetry of the Green function with respect to such an exchange) to obtain

$$G_N(r|r_0)\nabla^2\psi_N(r_0) - \psi_N(r_0)\nabla^2 G_N(r|r_0) = \delta(r-r_0)\psi_N(r_0) \qquad (3.4.10)$$

where ∇^2 now operates on the coordinates of r_0. Finally, both sides of equation (3.4.10) are integrated in the coordinates of r_0 over the region $\mathcal{R}$ of the real cavity but the volume integral of the left-hand side is simplified to an integral over the surface S by application of Green's theorem so that the result of this operation is

$$\psi_N(r) = \iint_S [G_N(r|r_S)(\partial/\partial n)\psi_N(r_S) - \psi_N(r_S)(\partial/\partial n)G_N(r|r_S)]\,dS$$

$$= -\iint_S \psi_N(r_S)[i(\omega/u)y(r_S)G_N(r|r_S) + (\partial/\partial n)G_N(r|r_S)]\,dS \qquad (3.4.11)$$

where r_0 has been specialized to r_S and $(\partial/\partial n)$ operates on the coordinates of r_S. This is the homogeneous integral equation that exactly specifies a solution of the homogeneous differential equation (3.4.6) that satisfies the boundary conditions of (3.4.7). Its solution will yield not only the form of the wavefunction but also the corresponding eigenvalues K_N; for if k does not take one of the values K_N the only possible solution, for free oscillations, is $\psi = 0$ everywhere inside $\Re$. Equation (3.4.11) is not yet in a form suitable for calculation; however, it will be convenient to continue the argument with separate consideration of the effects of non-zero surface admittance and of imperfect geometry. Ultimately we will introduce the restriction that both be small perturbations.

3.4.1   Non-zero surface admittance

We first consider the case of perfect geometry but non-zero effective specific acoustic admittance $y(r_S)$ of the surface. Since the surface S now coincides with S_0, the second term in the surface integral vanishes. The next step is to substitute equation (3.4.5), the series expansion of the Green function, and separate the dominant Nth term:

$$\psi_N(r) = -i\left(\frac{\omega}{u}\right) \iint\limits_S \psi_N(r_S)G'_N(r \,|\, r_S)y(r_S)\ \mathrm{d}S$$

$$-i\left(\frac{\omega}{u}\right)\left[\left(\iint\limits_S \psi_N(r_S)y(r_S)\phi^*_N(r_S)\ \mathrm{d}S\right)\left(V\Lambda^0_N(k^2_N - K^2_N)\right)^{-1}\right]\phi_n(r). \quad (3.4.12)$$

Here G'_N is the series for G_N neglecting the Nth term:

$$G'_N(r \,|\, r_0) = \sum_{M \neq N} \frac{\phi_M(r)\phi^*_M(r_0)}{V\Lambda^0_M(k^2_M - K^2_N)}. \quad (3.4.13)$$

Since equation (3.4.12) is homogeneous, we are at liberty to modify its solution by any factor a that is independent of r; we choose the one that makes the coefficient of $\phi_N(r)$ unity and that therefore satisfies

$$ia\left(\frac{\omega}{u}\right) \iint\limits_S \psi_N(r_S)y(r_S)\phi^*_N(r_S)\ \mathrm{d}S = -V\Lambda^0(k^2_N - K^2_N). \quad (3.4.14)$$

The particular solution that results from this operation

$$\Phi_N(r) = \phi_N(r) - i\left(\frac{\omega}{u}\right) \iint\limits_S \Phi_N(r_S)y(r_S)G'_N(r \,|\, r_S)\ \mathrm{d}S \quad (3.4.15)$$

has the property $\Phi_N(r) \rightarrow \phi_N(r)$ in the limit $y \rightarrow 0$. The corresponding eigenvalues may be obtained directly from equation (3.4.14) by recalling that $a\psi_N = \Phi_N$:

$$K^2_N = k^2_N + \left(\frac{i\omega}{uV\Lambda^0_N}\right) \iint\limits_S \Phi_N(r_S)y(r_S)\phi^*_N(r_S)\ \mathrm{d}S. \quad (3.4.16)$$

Equations (3.4.15) and (3.4.16) may be solved simultaneously to any desired degree of accuracy. For a first-order approximation we simply substitute ϕ_N for Φ_N inside the integrals to obtain, for the eigenvalue,

$$K_N = k_N + \left(\frac{\omega}{uk_N}\right)\left(\frac{i}{2V\Lambda_N^0}\right) \iint_S y(r_S)\,|\,\phi_N(r_S)\,|^2\,\mathrm{d}S \qquad (3.4.17)$$

correct to leading order in the surface admittance. Considerable use will be made of this result. For either free oscillations or forced oscillations at resonance, we may replace ω/uk_N by unity in a first-order approximation. Consequently, in the absence of losses other than at the surface, the Q for resonance of the Nth mode is given by

$$Q_N = \left[\left(\frac{\omega_N}{u}\right)V\Lambda_N^0\right]\left(\iint_S \xi(r_S)\,|\,\phi_N(r_S)\,|^2\,\mathrm{d}S\right)^{-1} \qquad (3.4.18)$$

where ξ is the specific acoustic conductance. This equation is consistent with the usual definition of the quality factor as the ratio of the energy stored by the standing wave to that dissipated per unit advance in the phase; the former is equal to the integral $\iiint_\mathfrak{R}(|\,p_a\,|^2/\rho u^2)\,\mathrm{d}V$ of the energy density over the volume of the cavity, while the latter is given by the surface integral $\iint_S[\xi(r_S)\,|\,p_a\,|^2/\rho u\omega]\,\mathrm{d}S$ of the boundary conductance $\xi(r_S)/\rho u$ multiplied by the magnitude of the acoustic pressure and divided by the angular frequency ω. Since the acoustic pressure is proportional to ϕ_N in our first-order approximation, and $\iiint_\mathfrak{R}|\phi_N|^2\,\mathrm{d}V = V\Lambda_N^0$, equation (3.4.18) follows directly.

3.4.2 Imperfect geometry

The case with imperfect geometry, with or without surface admittance, parallels the one outlined above and the exact results are

$$\Phi_N(r) = \phi_N(r) - \iint_S \Phi_N(r_S)\,[(\partial/\partial n)G_N'(r\,|\,r_S) + i(\omega/u)y(r_S)G_N'(r\,|\,r_S)]\,\mathrm{d}S \qquad (3.4.19)$$

and

$$K_N^2 = k_N^2 + \left(\frac{1}{V\Lambda_N^0}\right)\iint_S \Phi_N(r_S)\,[(\partial/\partial n)\phi_N^*(r_S) + i(\omega/u)y(r_S)\phi_N^*(r_S)]\,\mathrm{d}S \qquad (3.4.20)$$

where the perturbed surface S lies everywhere within or on S_0. When the distortion of the boundary shape is small, first-order approximations may be sufficient in which we set $\Phi_N(r_S) = \phi_N(r_S)$ and $K_N = k_N$ on the right-hand sides of equations (3.4.19) and (3.4.20). Specializing to the case of a cavity with zero wall admittance we then find

$$\Phi_N(r) = \phi_N(r) + \sum_{M \neq N} \frac{A_{NM}}{V\Lambda_M^0(k_N^2 - k_M^2)} \qquad (3.4.21)$$

and

$$K_N^2 = k_N^2 + (A_{NN}/V\Lambda_N^0) \qquad (3.4.22)$$

where A_{NM} is given by

$$A_{NM} = \iint\limits_{S} \phi_N(r_S)(\partial/\partial n)\phi_M^*(r_S) \, dS. \qquad (3.4.23)$$

Should the change in boundary shape be more substantial, an approximation of higher order may be required. Unfortunately, it turns out that the expansion of Φ_N in terms of the basis set ϕ_N does not converge very rapidly. This is because the functions Φ_N must vanish everywhere outside of the surface S, and therefore be discontinuous across it, while the functions ϕ_N are discontinuous only on S_0. However, it is possible to employ a basis set, formed by linear combinations of the functions ϕ_N, that does vanish on the new surface S. An improved perturbation treatment of that kind, described by Morse and Feshbach [2], can be used to obtain an expression for the eigenvalues correct to second order.

Another difficulty arises for degenerate modes because the corrections to the unperturbed wavefunctions then contain divergent terms with $k_M = k_N$. In that case it is necessary to employ a new basis set in which the s different wavefunctions ϕ_n ($n = 1, 2, \cdots, s$) that share the same eigenvalue are recast as s linear combinations

$$\psi_n = \sum_{r=1}^{s} [c_{nr}/(\Lambda_r^0)^{1/2}]\phi_r \qquad (3.4.24)$$

that each obey

$$\iint\limits_{S} \psi_n(r_S)(\partial/\partial n)\psi_m^*(r_S) \, dS = 0 \qquad n \neq m. \qquad (3.4.25)$$

When this new set of functions ψ_n is used in place of the original wavefunctions, the first-order expression for the perturbed eigenvalues becomes

$$K_n^2 = k_n^2 + \left(\iint\limits_{S} \psi_n(r_S)(\partial/\partial n)\psi_n^*(r_S) \, dS \right) \left(\iiint\limits_{\mathcal{R}_0} |\psi_n(r)|^2 \, dV \right)^{-1} \qquad (3.4.26)$$

and the divergent higher-order terms vanish. The coefficients c_{nr} in each linear combination may be identified with the elements of the corresponding eigenvector c_n of an $s \times s$ matrix B:

$$Bc_n = \lambda_n c_n. \tag{3.4.27}$$

The matrix elements B_{nm} must be proportional to surface integrals A_{nm} of equation (3.4.23) and we choose to write them as

$$B_{nm} = [(\Lambda_n^0 \Lambda_m^0)^{-1/2} / (2\varepsilon k_n^2 V)] A_{nm} \tag{3.4.28}$$

where ε is a small parameter that determines the magnitude of the shape distortion, so that the characteristic values λ_n are dimensionless quantities of order unity. Since in equation (3.4.26) the surface integral evaluates to $2\varepsilon k_n^2 V \lambda_n \Sigma_r c_{nr}^2$, and the volume integral is $V \Sigma_r c_{nr}^2$, the fractional frequency shifts may be obtained from the very simple formula

$$(K_n - k_n)/k_n = \varepsilon \lambda_n + O(\varepsilon^2). \tag{3.4.29}$$

The matrix B as defined above contains information about the change in both shape and volume. However, when we come to apply the method, it is often convenient to subtract from the diagonal elements of the matrix that part of the frequency shift due just to the change in volume so that we obtain results pertaining to changes in shape alone. An example of the application of boundary shape perturbation theory, for the case of the degenerate radial modes of a cylindrical resonator, will be given in §3.6.3.

3.5 The Parallelepiped

Now that we have a fully developed theory applicable to real resonators, we turn to the development of specific solutions. In each case we will first assume homogeneous Neumann boundary conditions and then apply the first-order perturbation theory to account for boundary layer effects and for the mechanical compliance of the walls; some plausible geometric imperfections will also be considered.

3.5.1 The ideal case

We begin with the simplest three-dimensional case: that of a parallelepiped, one corner of which we take as the origin, having sides of length L_x, L_y and L_z parallel to the rectangular Cartesian coordinates x, y and z. Substitution of the trial wavefunction $\phi_N = X_N(x) Y_N(y) Z_N(z)$ in the Helmholtz equation separates the problem into the solution of three ordinary differential equations:

$$\begin{aligned} (d^2 X/dx^2) + k_x^2 X &= 0 \\ (d^2 Y/dy^2) + k_y^2 Y &= 0 \\ (d^2 Z/dz^2) + k_z^2 Z &= 0 \end{aligned} \tag{3.5.1}$$

in which

$$k_x^2 + k_y^2 + k_z^2 = k^2. \tag{3.5.2}$$

The general solutions of these equations are combinations of $\sin(k_q q)$ and $\cos(k_q q)$, where q is a Cartesian coordinate, but, since the boundary conditions require the normal derivatives to vanish at the walls, the former are excluded and the k_q of the latter are restricted to integer multiples of π/L_q:

$$\phi_N(x, y, z) = \cos(n_x \pi x / L_x)\cos(n_y \pi y / L_y)\cos(n_z \pi z / L_z) \tag{3.5.3}$$

$$k_N^2 = (n_x \pi / L_x)^2 + (n_y \pi / L_y)^2 + (n_z \pi / L_z)^2 \tag{3.5.4}$$

$$\Lambda_N^0 = 1/\varepsilon_{n_x}\varepsilon_{n_y}\varepsilon_{n_z} \tag{3.5.5}$$

where $n_x, n_y, n_z = 0, 1, 2, \cdots, \varepsilon_0 = 1$ and $\varepsilon_{n>0} = 2$.

For each axis q with $n_q > 0$, there are $(n_q + 1)$ plane nodal surfaces perpendicular to the axis (including those at $q = 0$ and $q = L_q$). When only one of the indices n_q is non-zero, the mode is a purely longitudinal one consisting of a superposition of plane waves propagating back and forth parallel to the q axis. These nodes have resonance frequencies that are harmonics of $u/2L_q$.

3.5.2   Non-zero wall admittance

The effects of non-zero admittance of the boundary surface are computed from the first-order equation (3.4.17) which gives for the Nth mode at resonance:[4]

$$K_N = k_N + \left(\frac{i}{2V\Lambda_N^0}\right) \iint\limits_{S} \beta(r_S)\phi_N^2(r_S)\, dS. \tag{3.5.6}$$

Assuming that the effective specific acoustic admittance of the surface is constant over the face of each wall and that walls opposite each other have identical properties, this equation reduces to

$$K_N = k_N + i\left[\beta_x(\varepsilon_{n_x}/L_x) + \beta_y(\varepsilon_{n_y}/L_y) + \beta_z(\varepsilon_{n_z}/L_z)\right] \tag{3.5.7}$$

where β_x refers to the walls perpendicular to x etc. The contributions of the thermal and shear boundary layers to the specific admittance of the walls may be obtained from equations (2.4.30) and (2.4.31). Since the tangential component k_t of the propagation constant at a wall perpendicular to the q axis is $(k^2 - k_q^2)^{1/2}$, we have

$$\beta_q = (1 + i)(\omega/2u)[(\gamma - 1)\delta_h + (1 - k_q^2/k_N^2)\delta_s] + \beta_{q,m} \tag{3.5.8}$$

where $\beta_{q,m}$ is the mechanical contribution to the specific acoustic admittance of the wall.

3.5.3   Non-parallel walls

As an example of the application of first-order boundary shape perturbation theory, we compute the shift in the eigenvalue of a longitudinal
mode resulting from non-parallelism of the end walls. Let the axis in question be z and let the surface nominally at the end $z = L_z$ actually lie in the
plane $z = L_z - x \sin \varepsilon$ as shown in figure 3.2. This ensures that the perturbed surface lies on or within S_0. Since the unperturbed wavefunction is
just $\cos(n_z \pi z / L_z)$, the normal derivative $(\partial/\partial n)\phi_N$ is simply
$-(n_z \pi / L_z)\sin(n_z \pi z / L_z)$ in first order. Evaluating the integral A_{NN} in
equation (3.4.22) over the surface of the end we find

$$K_{0,0,n_z} = (n_z \pi / L_z)[1 + \tfrac{1}{2}\varepsilon(L_x/L_z) + O(\varepsilon^2)]. \qquad (3.5.9)$$

But the change in the boundary shape has also reduced the volume of the
cavity. If the change in the mean length, $-\tfrac{1}{2}\varepsilon L_x$ in first order, is restored
by uniform translation of the end then the fractional shift in the eigenvalue
is reduced to $O(\varepsilon^2)$. This is therefore an example of a shape perturbation,
the effect of which vanishes in first order when the volume is preserved.

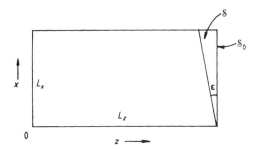

Figure 3.2 Resonator with non-parallel end walls. The perturbed
surface S is identical with S_0 everywhere except at one end which, while
still flat and parallel to the y axis, is skewed at an angle ε to the x axis.

3.6   The Cylindrical Cavity

Cylindrical polar coordinates (r, θ, z), related to the Cartesian set by

$$x = r \cos \theta \qquad y = r \sin \theta \qquad z = z \qquad (3.6.1)$$

will now be appropriate (see figure 3.3). In terms of these coordinates the
differential operators are given by

$$\nabla = e_r(\partial/\partial r) + e_\theta(1/r)(\partial/\partial \theta) + e_z(\partial/\partial z)$$
$$\nabla \cdot = (1/r)(\partial/\partial r)r e_r \cdot + (1/r)(\partial/\partial \theta)e_\theta \cdot + (\partial/\partial z)e_z \cdot \qquad (3.6.2)$$
$$\nabla^2 = (\partial^2/\partial r^2) + (1/r)(\partial/\partial r) + (1/r^2)(\partial^2/\partial \theta^2) + (\partial^2/\partial z^2)$$

where e_r, e_θ, e_z denote the unit vectors.

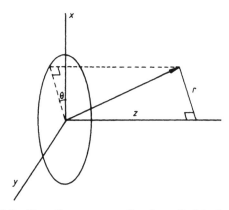

Figure 3.3   Coordinate system for the cylindrical cavity.

3.6.1   The ideal case

We consider a cylinder of radius b, concentric with the z axis, extending from $z = 0$ to $z = L$. The ideal wavefunctions for this cavity will be expressed as the product

$$\phi_N(r, \theta, z) = X_N(r) Y_N(\theta) Z_N(z). \tag{3.6.3}$$

When this is substituted into the wave equation, the problem separates into three ordinary differential equations

$$(1/X)(d^2 X/dr^2) + (1/rX)(d X/dr) + (k_N^2 - q^2) - (m/r)^2 = 0$$
$$(d^2 Y/d\theta^2) + m^2 Y = 0 \tag{3.6.4}$$
$$(d^2 Z/dz^2) + k_z^2 Z = 0$$

in which $-m^2$ is the separation constant for Y and $-k_z^2$ is the separation constant for Z. The substitution $r = x/(k_N^2 - k_x^2)^{1/2}$ reveals that the first of equations (3.6.4) is identical with Bessel's equation

$$(d^2 X/dx^2) + (1/x)(d X/dx) + [1 - (m/x)^2] = 0 \tag{3.6.5}$$

for which the solutions are $J_m(x)$, the cylindrical Bessel function of order m, and $N_m(x)$, the cylindrical Neumann function of order m. Since the Neumann functions are infinite along the line $r = 0$, part of which lies within the cavity, that solution must be rejected. The remaining solution must satisfy the boundary condition

$$(d/dr) J_m [r(k_N^2 - k_z^2)^{1/2}] \,|_{r=b} = 0 \tag{3.6.6}$$

which restricts $(k_N^2 - k_z^2)^{1/2} b$ to a discrete set of values χ_{mn}, the first few of which are given in table 3.1 [3, 4]. Some properties of the cylindrical Bessel functions of integer order are given in Appendix 1.

The solutions of the angular part of the wave equation are $Y^+ = \cos(m\theta)$ and $Y^- = \sin(m\theta)$, where the superscripts denote the symmetry of the function with respect to the sign of θ. Since the solutions are required to be periodic with period 2π, m is restricted to zero or integer values; negative integers are allowed but not required for a complete orthonormal set when the sine and cosine functions are taken separately. However, a rather more convenient solution is given with a linear combination of the two, $\cos(m\theta) + \sin(m\theta)$, in which m takes either sign: $m = 0, \pm 1, \pm 2, \cdots$.

The longitudinal part of the wave equation has the same solutions as in the case of the parallelepiped and again only a cosine function, $\cos(l\pi z/L)$, can satisfy the boundary conditions at both ends.

Table 3.1 Turning points of the cylindrical Bessel functions.

χ_{mn}	$n = 1$	2	3	4	5
$m = 0$	0.000 000	3.831 706	7.015 587	10.173 468	13.323 692
1	1.841 184	5.331 443	8.526 316	11.706 005	14.863 589
2	3.054 237	6.706 133	9.969 468	13.170 371	16.347 522
3	4.201 189	8.015 237	11.345 924	14.585 848	17.788 748
4	5.317 553	9.282 396	12.681 908	15.964 107	19.196 029

$\chi_{0n} \to \pi(n - \frac{3}{4})$, $n \gg 1$; $\chi_{m1} \to m + 0.809m^{1/3}$, $m \gg 0$;
$\chi_{mn} \to \pi(n + \frac{1}{2}m - \frac{3}{4}) - [(4m^2 - 3)/8\pi(n + \frac{1}{2}m - \frac{3}{4})]$, m fixed, $n \gg 1$.

The full solution for the ideal cylindrical cavity is therefore given by:

$$\phi_N(r, \theta, z) = J_m(\chi_{mn}r/b)[\cos(m\theta) + \sin(m\theta)]\cos(l\pi z/L) \quad (3.6.7)$$

$$k_N^2 = (l\pi/L)^2 + (\chi_{mn}/b)^2 \quad (3.6.8)$$

$$l, |m| = 0, 1, 2, \cdots, \qquad J_{-m}(x) = (-1)^m J_m(x).$$

Each value of $|m|$, other than zero, corresponds to a pair of degenerate modes, one for $+|m|$ and one for $-|m|$. Using appropriate properties of the Bessel and trigonometric functions (see Appendix 1), the orthogonality of the wavefunctions can be verified and the normalization constants shown to be

$$\Lambda_N^0 = (1/\varepsilon_l)[1 - (m/\chi_{mn})^2]J_m^2(\chi_{mn}) \quad (3.6.9)$$

where $\varepsilon_0 = 1$ and $\varepsilon_{l>0} = 2$.

The Nth mode has $(l + 1)$ plane nodal surfaces ($l > 0$) perpendicular to the axis, $|m|$ plane radial nodal surfaces, and n cylindrical nodal surfaces concentric with the axis (except for $m = 0$ when there are $n - 1$ cylindrical nodes[6]). The shapes of the characteristic functions for modes of various symmetries are shown in figure 3.4. The purely longitudinal modes (figure 3.4(a)), involving plane waves reflecting back and forth between the end plates, are those for which $n = 1$ and $m = 0$; they have characteristic functions $\cos(l\pi z/L)$ and resonance frequencies that are harmonics of $u/2L$. Purely radial modes (figure 3.4(b)), $l = m = 0$, have the characteristic functions $J_0(\chi_{0n}r/b)$ and resonance frequencies $\chi_{0n}(u/2\pi b)$ that are not harmonics of the fundamental. These modes tend to concentrate the acoustic energy along the axis of the cylinder, rather as one might expect for waves focused in on the centre by the cylindrical surface. Compound modes with both longitudinal and radial motion excited, but still with $m = 0$, preserve the axial symmetry of the pure radial and longitudinal modes and have characteristic functions that resemble figure 3.4(a) along the axis and figure 3.4(b) along radii. This symmetry is broken when $|m| \neq 0$. The azimuthal modes with $n = 1$ tend to concentrate the energy near the wall (figure 3.4(c)) with the fluid motion primarily that of circulation back and forth around the axis, while modes with both radial and azimuthal excitation distribute the energy with increasing uniformity as the indices increase (figure 3.4(d)).

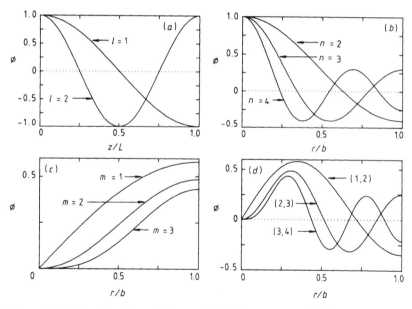

Figure 3.4 Wavefunctions ϕ for a cylindrical resonator: (a) for pure longitudinal modes ($n = 1, m = 0$); (b) for pure radial modes ($l = 0, m = 0$); (c) for azimuthal modes ($n = 1, l = 0$); (d) for radial/azimuthal compound modes ($l = 0$) labelled (m, n).

3.6.2 Non-zero wall admittance

The complex natural frequencies F_N of the resonator are given in terms of the eigenvalues by equation (3.2.8). When, as above, the surface admittance is taken as zero, the eigenvalues are purely real, the resonance frequencies f_N equal $(u/2\pi)k_N$, and the only term contributing to the resonance half-width g_N is that due to the bulk absorption: $g_b = (u/2\pi)\alpha$. Both the real and imaginary parts of the eigenvalues will be modified by non-zero surface admittance but, under most circumstances, the shifts can be accounted for with sufficient accuracy using the first-order theory as expressed in equation (3.4.17). The surface integral appearing in that expression may be written as the sum of integrals over the sides and the two ends of the cylinder:

$$\iint_S y(r_S)\phi_N^2(r_S)\ \mathrm{d}S = \int_0^L \int_0^{2\pi} y(r = b, \theta, z)\phi_N^2(r = b, \theta, z)b\ \mathrm{d}\theta\ \mathrm{d}z$$

$$+ \int_0^b \int_0^{2\pi} y(r, \theta, z = 0)\phi_N^2(r, \theta, z = 0)\ \mathrm{d}\theta\ r\mathrm{d}r$$

$$+ \int_0^b \int_0^{2\pi} y(r, \theta, z = L)\phi_N^2(r, \theta, z = L)\ \mathrm{d}\theta\ r\mathrm{d}r$$

$$= (V/\varepsilon_l)J_m^2(\chi_{mn})$$

$$\times \{(2y_{\text{side}}/b) + (2\varepsilon_l y_{\text{end}}/L)[1 - (m/\chi_{mn})^2]\}. \quad (3.6.10)$$

Here y_{side} and y_{end} are the specific acoustic admittances of the side and the ends respectively. For the present, we shall ignore mechanical terms in the surface admittance and concentrate on those due to the boundary layers, which, ignoring molecular slip and the temperature jump, may be computed from equations (2.4.30) and (2.4.31). The use of those formulae, derived in connection with plane-wave reflection, to describe the viscous and thermal effects at the cylindrical surface of the cavity is valid as long as $\delta_s, \delta_h \ll b$. Typically, the boundary layers are less than a few tenths of a millimetre thick so that these conditions are usually readily met in practice. Since the viscous damping at the walls depends upon the tangential component k_t of the propagation constant, the boundary admittance will depend upon the mode of the cavity's excitation. In leading order, k_t^2 is equal to $(l\pi/L)^2 + (m/b)^2$ along the side and $(\chi_{mn}/b)^2$ at the ends, so that the effective specific acoustic admittances are given by

$$y_{\text{side}} = (1 + \mathrm{i})\left(\frac{\omega}{2u}\right)\left((\gamma - 1)\delta_h + \frac{(l\pi/L)^2 + (m/b)^2}{(l\pi/L)^2 + (\chi_{mn}/b)^2}\delta_s\right) \quad (3.6.11)$$

and

$$y_{\text{end}} = (1 + \mathrm{i})\left(\frac{\omega}{2u}\right)\left((\gamma - 1)\delta_h + \frac{(\chi_{mn}/b)^2}{(l\pi/L)^2 + (\chi_{mn}/b)^2}\delta_s\right) \quad (3.6.12)$$

where $\delta_h = (2D_h/\omega)^{1/2}$ and $\delta_s = (2D_s/\omega)^{1/2}$. Thus, ignoring coupling between fluid and shell motion but otherwise taking the most general case, the resonance frequencies and half-widths correct to first order are

$$f_N = (u/2\pi)[(l\pi/L)^2 + (\chi_{mn}/b)^2]^{1/2} + \Delta f_s + \Delta f_h \qquad (3.6.13)$$

and

$$g_N = g_s + g_h + g_b. \qquad (3.6.14)$$

In these equations, Δf_s and Δf_h (both negative) are shifts from the unperturbed resonance frequencies due to the viscous and thermal boundary layers, while g_s and g_h are the corresponding contributions to the half-widths. Combining equation (3.4.17) with (3.6.9–3.6.12), we have

$$g_h = -\Delta f_h = \frac{f(\gamma - 1)\delta_h}{2b[1 - (m/\chi_{mn})^2]} \, [(1 + \varepsilon_l R) - (m/\chi_{mn})^2 \varepsilon_l R] \qquad (3.6.15)$$

and

$$g_s = -\Delta f_s = \frac{f\delta_s}{2b[1 - (m/\chi_{mn})^2]} \left(1 + \frac{(1 - \varepsilon_l R)(m^2 - \chi_{mn}^2)}{(l\pi R)^2 + \chi_{mn}^2}\right) \qquad (3.6.16)$$

where $R = b/L$.

The modes with $m > 0$ suffer a much greater damping than the axially symmetric modes ($m = 0$) because the wave energy is held close to the wall and is therefore dissipated more rapidly in the boundary layers. For a given value of m, this dissipation is most severe for the mode with $n = 1$. Indeed, for large values of m the ratio m/χ_{m0} approaches unity and the Q for these modes approaches zero. Usually, however, we are more interested in axially symmetric modes for which a simpler expression for the boundary layer contributions to the half-widths may be given:

$$2(g_h + g_s)/f = (\gamma - 1)(\delta_h/b)(1 + \varepsilon_l R) + (\delta_s/b)[1 + (\varepsilon_l R - 1)\beta_{ln}^2] \qquad (3.6.17)$$

where

$$\beta_{ln}^2 = \chi_{0n}^2/[(l\pi R)^2 + \chi_{0n}^2]. \qquad (3.6.18)$$

In the case of purely longitudinal excitation, n is unity and we have

$$2(g_h + g_s)/f = (\gamma - 1)(\delta_h/b)(1 + 2R) + (\delta_s/b)$$
$$= (\gamma - 1)(\pi f b^2)^{-1/2}(1 + 2R)D_h^{1/2} + (\pi f b^2)^{-1/2}D_s^{1/2} \qquad (3.6.19)$$

while for purely radial modes, l is zero and

$$2(g_h + g_s)/f = (\gamma - 1)(\delta_h/b)(1 + R) + (\delta_s/L)$$
$$= (\gamma - 1)(\pi f b^2)^{-1/2}(1 + R)D_h^{1/2} + (\pi f L^2)^{-1/2}D_s^{1/2}. \qquad (3.6.20)$$

The boundary layer theory, while a good approximation, is not exact. In gases at low pressures, the effects of molecular slip and the temperature jump at the walls of the resonator need to be included and the surface admittance should then be calculated from equation (2.4.45) and (2.4.46). The half-widths remain unchanged by inclusion of these effects but the frequency shifts are modified slightly to become

$$-\Delta f_h = g_h(1 - 2l_h/\delta_h) \qquad (3.6.21)$$

$$-\Delta f_s = g_s(1 - 2l_s/\delta_s) \qquad (3.6.22)$$

where l_h and l_s are the thermal and shear accommodation lengths given by equations (2.4.44) and (2.4.41). Also, for narrow tubes, δ_s and δ_h may no longer be considered small compared with the radius; indeed the boundary layers may merge, and a more accurate solution of the boundary value problem is required with the thermal and shear modes developed explicitly in cylindrical coordinates [14,15].

The contribution of the wall itself to the effective surface admittance may not always be neglected. Unfortunately, there is no exact solution of the problem in closed form and detailed approximations do not appear to have been worked out. However, one can infer from general considerations that the coupling of fluid and shell motion must cause a perturbation Δf_{sh} to the resonance frequencies that is linear in the fluid density. Unless the frequency happens to be coincident with a resonance of the shell, the effect is expected to be small (and comparable with that discussed in §3.7.2 for a spherical resonator).

As an indication of the magnitude of the coupling, consider the speed u_{01} and absorption coefficient α_{01} of a nominally plane travelling wave in an infinite cylindrical waveguide with yielding walls. The acoustic pressure of the travelling wave must of course be a solution of the Helmholtz equation and so we write

$$p_a(r, z) = AJ_0(k_r r)\exp(-ik_{01}z) \qquad (3.6.23)$$

with

$$k_r = (k^2 - k_{01}^2)^{1/2}. \qquad (3.6.24)$$

In equation (3.6.23), A is a constant and $k_{01} = (\omega/u_{01}) - i\alpha_{01}$ is the effective propagation constant for the plane wave travelling along the tube in the positive z direction. The value of k_r differs from zero only as a result of the non-zero boundary admittance and its value is determined by satisfying equation (2.4.29) on the surface $r = b$. Making use of standard properties of the Bessel functions (see Appendix 1), it is easy to show that

$$(1/p_a)(\partial p_a/\partial r)_{r=b} = -k_r[J_1(k_r b)/J_0(k_r b)]$$
$$= -k_r[(k_r b/2) + \tfrac{1}{2}(k_r b/2)^3 + \cdots]. \qquad (3.6.25)$$

Consequently, neglecting the bulk absorption, the boundary conditions require that

$$k_{01} = (\omega/u) - (i/b)y_s \qquad (3.6.26)$$

correct to leading order in the specific acoustic admittance of the surface,

y_S. At frequencies small compared with that of the lowest radial resonance of the wall, this admittance is given by

$$y_S = -i\rho u\omega bC \qquad (3.6.27)$$

where $C = (1/b)(\partial r/\partial p)$ is the static compliance of the tube with respect to variation of the internal pressure. The ratio u_{01}/u is therefore given correct to leading order by

$$u_{01}/u = 1 - \rho u^2 C. \qquad (3.6.28)$$

Since y_S is purely imaginary in this approximation, there is no contribution to the effective absorption coefficient. Typically, C is of order 10^{-10} Pa^{-1} or less for the sort of tubes used in the fabrication of cylindrical resonators[5] and the effect of wall motion on the observed sound speed in gases (where $\rho u^2 \sim \gamma p$ at low pressures) is very small. For a liquid-filled tube, the effect could be rather larger ($\rho u^2 \approx 2.3$ GPa in liquid water) but liquids are usually studied at high frequencies where equation (3.6.27) is inapplicable and the motion is controlled by the mass of the tube.

A further complication arises in a closed cylindrical resonator for all modes with odd values of the longitudinal index l; these modes exert forces of opposite sign on each end plate and therefore couple to translation of the resonator. A simple analysis for the case in which the shell is free indicates that $y_{end} = -(i/kL)(m_f/m_s)$, where m_f and m_s are the total mass of the fluid and shell respectively, so that there is a fractional increase in the odd-ordered longitudinal resonance frequencies of $[2/(l\pi)^2](m_f/m_s)$. Practical methods of holding the resonator may be expected to modify this behaviour somewhat and introduce some damping.

Incidentally, the treatment given above can also be used to derive the Kirchhoff–Helmholtz value of the tube propagation constant. If we consider only the viscothermal effects at the wall, neglecting both the bulk absorption and mechanical terms in the surface admittance, then equation (3.6.26) applies with $y_S = (1 + i)(\omega/2u)[\delta_s + (\gamma - 1)\delta_h]$, from (2.4.30) and (2.4.31), and (2.4.35) follows directly.

3.6.3 Imperfect geometry

The modes of the cylindrical resonator most frequently employed in practice are the longitudinal and the radial modes. It is straightforward to show that these modes are insensitive to volume-preserving geometric imperfections and, in particular, that the first-order frequency shifts due to non-parallelism of the end plates and to eccentricity of the bore vanish.

As an example of the application of the general first-order treatment for degenerate modes, we shall consider the azimuthal modes ($l = 0$; $n, |m| \neq 0$) where each value of $|m|$ corresponds to a degenerate pair in a cylindrical resonator of perfect geometry. It is convenient now to work with a complex form of the unperturbed wavefunctions which, with the corresponding normalization factors, we write as

$$\phi_{nm} = J_{|m|}(\chi_{mn}r/b)\exp(im\theta) \qquad (3.6.29)$$

$$V\Lambda^0_{nm} = \pi b^2 L\,[1 - (m/\chi_{mn})^2]\,J^2_{|m|}(\chi_{mn}). \qquad (3.6.30)$$

We shall consider a distortion of the cylindrical wall, independent of z, such that the surface is given by

$$r = b\,[1 - \varepsilon f(\theta)] \qquad (3.6.31)$$

where ε is a small parameter and $f(\theta)$ is a function of order unity that we expanded as a Fourier series

$$f(\theta) = \sum_{s=-\infty}^{\infty} C_s \exp(is\theta) \qquad (3.6.32)$$

the coefficients of which must satisfy $C_s^* = C_{-s}$ so that $f(\theta)$ is real. In terms of this function, the normal derivative which appears in the perturbation expressions is equal to $(e_r + \varepsilon b\nabla f)\cdot\nabla$. In formulating the elements of the matrix B we need consider only the two wavefunctions with the same values of n and $|m|$ for which in leading order we have

$$\frac{\phi^*_{nm'}(\partial/\partial n)\phi_{nm}}{2\varepsilon k^2_{nm}V\Lambda^0_{nm}}$$

$$= \sum_{s=-\infty}^{\infty} C_s\left(\frac{1 - (m/\chi_{nm})^2 - (sm/\chi^2_{nm})}{2\pi bL\,[1 - (m/\chi_{nm})^2]}\right)\exp\,[i(m - m' + s)\theta] \qquad (3.6.33)$$

with $|m'| = |m|$. To obtain the matrix elements $B_{m'm}$, we integrate equation (3.6.33) over the surface and subtract from the diagonal elements the fractional shift in the resonance frequency due to the change in volume alone:

$$B_{m'm} = \int_0^L\int_0^{2\pi} \frac{\phi^*_{nm'}(\partial/\partial n)\phi_{nm}}{2\varepsilon k^2_{nm}V\Lambda^0_{nm}}\,b\,\,d\theta\,\,dz - C_0\delta_{m'm}. \qquad (3.6.34)$$

Since the only non-zero terms in these integrals are those with $(m - m' + s) = 0$, the matrix is given by

$$B = \left(\frac{1 + (m/\chi_{nm})^2}{1 - (m/\chi_{nm})^2}\right)\begin{pmatrix} 0 & C_{2m}^* \\ C_{2m} & 0 \end{pmatrix} \qquad (3.6.35)$$

and the characteristic values are

$$\lambda_{\pm m} = \pm\,|C_{2m}|\left(\frac{1 + (m/\chi_{nm})^2}{1 - (m/\chi_{nm})^2}\right). \qquad (3.6.36)$$

Consequently, the effect of the change in shape is to split the two resonance frequencies symmetrically about those for the perfect cylinder by an amount proportional to $\varepsilon|C_{2m}|$. The average of the two frequency shifts is seen to vanish in the first order of perturbation theory. When C_{2m} is real, the only important component of $f(\theta)$ is $\cos(2m\theta)$ and the wavefunctions determined by the linear combinations are are

$$\psi_{n,+m} \propto J_m(\chi_{mn}r/b)\cos(m\theta) \tag{3.6.37}$$

$$\psi_{n,-m} \propto J_m(\chi_{mn}r/b)\sin(m\theta). \tag{3.6.38}$$

3.7   The Spherical Enclosure

Spherical polar coordinates (r, θ, ξ) are now the most appropriate ones with which to proceed (see figure 3.5). In this system, r is the magnitude of the position vector, the co-latitude θ is the angle made with the Cartesian z axis, and the azimuthal angle ξ is made between the projection of the position vector on the xy plane and the x axis:

$$x = r \sin \theta \cos \xi \qquad y = r \sin \theta \sin \xi \qquad z = r \cos \theta. \tag{3.7.1}$$

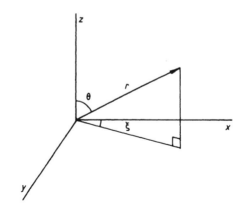

Figure 3.5   Coordinate system for the spherical cavity.

The differential operators are therefore given by

$$\nabla = e_r(\partial/\partial r) + e_\theta(1/r)(\partial/\partial\theta) + e_\xi[1/(r \sin \theta)](\partial/\partial\xi)$$
$$\nabla \cdot = [(1/r^2)(\partial/\partial r)r^2 e_r \cdot] + [1/(r \sin \theta)][(\partial/\partial\theta)(\sin \theta)e_\theta \cdot]$$
$$+ [1/(r \sin \theta)][(\partial/\partial\xi)e_\xi \cdot] \tag{3.7.2}$$
$$\nabla^2 = (\partial^2/\partial r^2) + (2/r)(\partial/\partial r) + (1/r^2)(\partial^2/\partial\theta^2)$$
$$+ [\cos \theta/(r^2 \sin \theta)](\partial/\partial\theta) + [1/(r^2 \sin^2 \theta)](\partial^2/\partial\xi^2)$$

where e_r, e_θ and e_ξ are the unit base vectors.

3.7.1 The ideal case

The region of the cavity is defined by a sphere of radius a centred at the origin. Perfect geometry and zero surface admittance are both assumed initially and the wavefunctions are separated into the product

$$\phi_N(r, \theta, \xi) = R_N(r) P_N(\theta) Q_N(\xi) \tag{3.7.3}$$

which, when substituted in the Helmholtz equation, leads to the three independent equations

$$(1/R)(d^2R/dr^2) + (2/rR)(dR/dr) + k^2 - l(l+1)/r^2 = 0$$
$$(d^2P/d\theta^2) + (\cos\theta/\sin\theta)(dP/d\theta) + [l(l+1) - m^2/\sin^2\theta] P = 0 \tag{3.7.4}$$
$$(d^2Q/d\xi^2) + m^2Q = 0.$$

Here, $-m^2$ is the separation constant for $Q(\xi)$ and $-l(l+1)$ is the separation constant for $P(\theta)$. The equation for the azimuthal factor Q is the same as in the cylindrical case and again the most convenient solution is

$$Q(\xi) = \cos(m\xi) + \sin(m\xi) \tag{3.7.5}$$

with m equal to a signed integer. The solution of the second of equations (3.7.4) is a little more complicated. When m is zero, an acceptable series solution in powers of $\eta = \cos\theta$ may be developed; this solution is called the lth Legendre polynomial of the first kind and is denoted by $P_l(\eta)$. In order that the solution remain finite at the poles of the sphere ($\cos\theta = \pm 1$), l is restricted to positive integer values or zero: $l = 0, 1, 2, \cdots$. The first few solutions are

$$P_0(\eta) = 1 \qquad\qquad P_0(\cos\theta) = 1$$
$$P_1(\eta) = \eta \qquad\qquad P_1(\cos\theta) = \cos\theta$$
$$P_2(\eta) = \tfrac{1}{2}(3\eta^2 - 1) \qquad P_2(\cos\theta) = \tfrac{1}{4}(3\cos 2\theta + 1).$$

Having found the solution for this special case it is a simple matter to show that the solution for values of m other than zero is

$$P_l^{|m|}(\eta) = (1 - \eta^2)^{|m|/2}(d^{|m|}/d\eta^{|m|}) P_l(\eta). \tag{3.7.6}$$

These functions $P_l^{|m|}(\eta)$ are called associated Legendre polynomials of the first kind; since they vanish for $|m| > l$, we need consider values of the integer m only in the range $-l$ to $+l$. The solutions for the two angular equations may be combined to define the set of spherical harmonics $Y_{lm}(\theta, \xi)$:

$$Y_{lm}(\theta, \xi) = [\cos(m\xi) + \sin(m\xi)] P_l^{|m|}(\cos\theta) \tag{3.7.7}$$

where $l = 0, 1, 2, \cdots$, and $m = 0, \pm 1, \pm 2, \cdots, \pm l$. Important properties of the Legendre functions and spherical harmonics are reviewed in Appendix 1.

For the radial factor the substitution $x = rk$ reveals that the first of equations (3.7.4) is

$$(d^2R/dx^2) + (2/x)(dR/dx) + [1 - l(l+1)/x^2]R = 0 \qquad (3.7.8)$$

for which the solution that is finite at the origin is $R = j_l(x)$, the spherical Bessel function of order l. The set of eigenvalues k_N then follows from the equation for the radial boundary condition

$$(d/dr)j_l(kr)|_{r=a} = 0 \qquad (3.7.9)$$

successive roots of which, $k_N a = \nu_{ln}$, are labelled by $n = 1, 2, 3, \cdots$. The first few values of ν_{ln} are given in table 3.2 and a comprehensive list can be found in the literature [3].

Table 3.2 Turning points of the spherical Bessel functions.

ν_{ln}	$n = 1$	2	3	4	5
$l = 0$	0	4.493 409	7.725 252	10.904 12	14.066 19
1	2.081 576	5.940 370	9.205 840	12.404 45	15.579 24
2	3.342 094	7.289 932	10.613 86	13.846 11	17.042 90
3	4.514 100	8.583 755	11.972 73	15.244 51	18.468 15
4	5.646 703	9.840 446	13.295 56	16.609 35	19.862 42

$\nu_{0n} \to \pi(n - \frac{1}{2})$, $n \gg 1$; $\nu_{l1} \to (l + \frac{1}{2}) + 0.809(l + \frac{1}{2})^{1/3}$, $l \gg 0$;
$\nu_{ln} \to \pi(n + \frac{1}{2}l - \frac{1}{2}) - \{[(2l + 1)^2 + 7]/8\pi(n + \frac{1}{2}l - \frac{1}{2})\}$, l fixed, $n \gg 1$.

The full solution for the ideal spherical cavity is therefore given by

$$\phi_N(r, \theta, \xi) = j_l(\nu_{ln}r/a)Y_{lm}(\theta, \xi) \qquad (3.7.10)$$

$$k_N = \nu_{ln}/a \qquad (3.7.11)$$

where $l = 0, 1, 2, \cdots$, $|m| = 0, \pm 1, \pm 2, \cdots, \pm l$, and $n = 1, 2, 3, \cdots$; the states with a given l are thus $(2l + 1)$-fold degenerate. Using the appropriate properties of the functions j_l and Y_{lm}, the orthogonality of the set of normal modes may be verified and the normalization constants shown to be

$$\Lambda_N^0 = \frac{3}{2} \left(\frac{1 - l(l+1)/\nu_{ln}^2}{(2l+1)} \right) \left(\frac{(l + |m|)!}{(l - |m|)!} \right) j_l^2(\nu_{ln}). \qquad (3.7.12)$$

In this ideal case, the resonance frequencies are just $f_{ln} = \nu_{ln}(u/2\pi a)$ and the only contribution to the half-widths g_{ln} is the bulk absorption term $g_b = (u/2\pi)\alpha$.

The Nth mode has l conical nodal surfaces, $|m|$ plane nodal surfaces perpendicular to the 'equatorial' plane $\theta = \pi/2$, and n spherical nodal surfaces (except for $l = 0$ when there are $n - 1$ spherical nodes[6]). The shapes of the characteristic functions for modes of various symmetries are shown in figure 3.6. The purely radial modes (figure 3.6(a)) resemble those of the cylinder and tend to concentrate the acoustic energy near to the centre of the sphere. These are the only non-degenerate modes in a perfect sphere. The modes with $l > 0$ but $m = 0$ retain axial symmetry and tend to concentrate the energy near to the wall when $l > n$ (figure 3.6(b)). When both l and n are large, the wave energy is distributed, on average, more evenly throughout the sphere (figures 3.6(c) and (d)). Modes with given values of l and n are characterized by the same total angular momentum and by the same radial factor in the wavefunction; they differ, through the value of m, in the alignment of the angular momentum vector in space. This behaviour closely resembles that of the hydrogen atom in quantum mechanics, where the angular factors in the wavefunctions are also spherical harmonics.

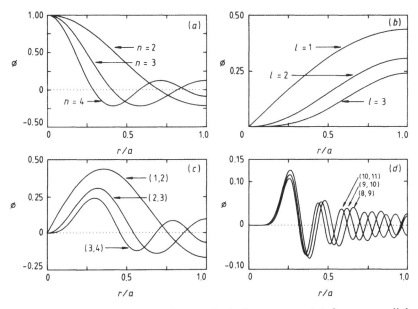

Figure 3.6 Wavefunctions ϕ for a spherical resonator: (a) for pure radial modes ($l = 0$, $n > 1$); (b) non-radial modes with $n = 1$; (c) and (d) compound modes labelled (l, n).

3.7.2 *Non-zero wall admittance* [6]

The case of the radially symmetric modes is particularly simple. The admittance is uniform over the whole of the surface and, since tangential motion is absent, there is no viscous damping at the wall. For these modes the effective specific acoustic admittance of the surface is simply

$$y_S = (1 + i)(\omega/2u)(\gamma - 1)\delta_h + y_{sh} \tag{3.7.13}$$

where y_{sh} is the admittance of the shell. The surface integral appearing in the equation for the perturbed eigenvalues, equation (3.4.17), is

$$\iint\limits_S y(r_S)\phi_N^2(r_S) \, dS = y_S a^2 j_0^2(\nu_{0n}) \int_0^{2\pi} d\xi \int_0^{\pi} \sin\theta \, d\theta$$

$$= 4\pi a^2 y_S j_0^2(\nu_{0n}) \tag{3.7.14}$$

and $V\Lambda_{0n} = 2\pi a^3 j_0^2(\nu_{0n})$. Consequently, the first-order expression for the resonance frequencies and half-widths of the radial modes is

$$f_{0n} + i g_{0n} = (u/2\pi a)(\nu_{0n} + i y_S + i a\alpha)$$

$$= (u/2\pi a)\nu_{0n} + (\Delta f_h + \Delta f_{sh}) + i(g_h + g_{sh} + g_b). \tag{3.7.15}$$

Here, Δf_h and Δf_{sh} are the shifts in the resonance frequency from the unperturbed value $(u/2\pi a)\nu_{0n}$ arising from the thermal boundary layer and the coupling of fluid and shell motion, while g_h and g_{sh} are the corresponding contributions to the half-width. From equation (2.4.45), which includes the effects of imperfect thermal accommodation, the boundary layer terms are found to be

$$-\Delta f_h = g_h(1 - 2l_h/\delta_h) \tag{3.7.16}$$

and

$$g_h = (\gamma - 1)(f/2a)\delta_h \tag{3.7.17}$$

where the accommodation length l_h is given by (2.4.44) and $\delta_h = (2D_h/\omega)^{1/2}$.

Calculation of the mechanical admittance of a resonator's wall usually involves many approximations. However, for an isotropic spherical shell, the admittance is amenable to exact calculation. This problem has been examined by Mehl [5] and, with inclusion of losses in the wall material and radiation from the outer surface, by Moldover *et al* [6]. Usually the effect of the wall motion is small and a simplification of the exact results will suffice:

$$y_{sh} = i2\pi f\rho u a C/ [1 - (f/f_0)^2]. \tag{3.7.18}$$

In equation (3.7.18), C is the static compliance of the shell for variation of the internal pressure and is given by

$$C = r^{-1}\left(\frac{\partial r}{\partial p}\right) = \frac{1 - \sigma}{2[(b/a)^3 - 1]\rho_w u_w^2}\left(\frac{(b/a)^3}{1 - 2\sigma} + \frac{2}{1 + \sigma}\right) \tag{3.7.19}$$

while f_0 is the frequency of the lowest radially symmetric resonance of the shell and is given approximately by

$$f_0 = \left(\frac{u_w}{2\pi a}\right) \left(\frac{2\,[(b/a)^3 - 1]}{[(b/a) - 1]\,[1 + 2(b/a)^3]}\right)^{1/2}. \tag{3.7.20}$$

Here, b is the outer radius of the spherical shell, and σ, ρ_w and u_w denote Poisson's ratio, density and longitudinal sound speed for the wall material respectively. The coupling of fluid and shell motion is thus linear in the fluid density and usually small for gases when the frequency is below the fundamental 'breathing' resonance of the shell. Note that iy_{sh} is purely real in this approximation and therefore contributes only a shift in the resonance frequencies of the fluid within the cavity,

$$\Delta f_{sh} = -f\rho u^2 C/\,[1 - (f/f_0)^2] \tag{3.7.21}$$

without a corresponding contribution to the half-width. Although losses due to radiation from the external surface can also be incorporated in the exact theory of the shell response, an approximation will suffice. We include in series with the lossless impedance of the shell an additional acoustical resistance R_a leading to

$$y_{sh} = i2\pi f\rho uaC/\,[1 - (f/f_0)^2 + i(f/f_0 Q_{sh})] \tag{3.7.22}$$

where $Q_{sh} = (2\pi f_0 aCR_a)^{-1}$ is a quality factor for the damped shell resonance. Since the speeds of the internal and external surfaces of the shell are approximately equal at frequencies small compared with $u_w/(b - a)$, and the ratio of external and internal surface areas is $(b/a)^2$, radiation into the external medium will be expected to contribute $(b/a)^2 \rho_0 u_0$ to the acoustic resistance at $r = a$, where $\rho_0 u_0$ is the characteristic impedance of the external fluid. For a typical resonator operating at a frequency well below the breathing resonance of the shell, this loss term will be negligible when the external fluid is a gas at a pressure of order 1 MPa or less but if it were a liquid (or a dense gas) then the quality factor would be small and g_{sh} may be comparable with Δf_{sh} [6]. The quality factor can also be adjusted to account for losses in the wall material itself [6].

Should there be an opening in the wall of the resonator, then the specific acoustic admittance will differ from that given by equation (3.7.13) over the area ΔS of that opening. In leading order, the shift Δf_0 in the resonance frequency and contribution g_0 to the half-width of a radial mode caused by such an opening will be

$$\Delta f_0 + ig_0 = (u/2\pi a)(\Delta S/4\pi a^2)iy_0 \tag{3.7.23}$$

where y_0 is the specific acoustic input admittance of the opening. The case of a tubular opening will be considered in Chapter 5.

For the non-radial modes, viscous damping at the wall must be considered. Since the ratio $(k_t/k)^2$ (where k_t is the tangential propagation constant) is equal to $l(l+1)v_{ln}^{-2}$ at the surface, the effective specific acoustic admittance, ignoring molecular slip and the temperature jump, becomes

$$y_S = (1 + i)(\omega/2u)\{(\gamma - 1)\delta_h + [l(l+1)v_{ln}^{-2}]\delta_s\} + y_m \tag{3.7.24}$$

and is again independent of angular coordinates except possibly through the mechanical term. This may then be used in the expression for the resonance frequencies and half-widths which, generalized to non-radial modes, is

$$f_{ln} + \mathrm{i}g_{ln} = (u/2\pi a)[\![\, v_{ln} + \mathrm{i}\{y_{\mathrm{S}}/[1 - l(l+1)v_{ln}^{-2}]\} + \mathrm{i}a\alpha\,]\!] \quad (3.7.25)$$

for y_{S} constant on S. Small corrections for openings in the wall can also be derived but, before they could be applied, it would be necessary to establish the location in space of the axis of the non-radial mode.

The exact theory of classical elasticity is soluble for the general case of non-radial motion [5]. However, the results are much more complicated than for the $l = 0$ case and are not readily reduced to simple expressions. The general behaviour for non-radial modes is, however, similar to the $l = 0$ case: the effects of coupling between fluid and shell motion are linear in the fluid density; and for dilute gases resonating at frequencies remote from those of shell resonances, the frequency shifts are small and damping is absent. However, there are several non-radial shell resonances that occur at frequencies well below the breathing mode and these can couple strongly to non-radial resonances of the fluid. In addition, for the $l = 1$ modes only, the fluid exerts a net force on the shell and thereby excites translational motion [6] (this is similar to the case of odd-order longitudinal modes in a cylindrical resonator).

It is worth remarking that an alternative treatment of the whole problem becomes attractive when one considers fluids of liquid density contained in thin-walled spherical shells because the specific acoustic admittance is then very large. The alternative approach starts with an ideal case in which a pressure-release boundary condition is assumed: $j_l(ka) = 0$; the allowed values of ka then correspond to the zeros of j_l rather than to its turning values. Thermal and viscous boundary layer effects are absent in this limit and small corrections for the departure of the wall impedance from zero can be estimated. This will be considered further in Chapter 7.

3.7.3 Imperfect geometry

The effects of imperfect geometry on the resonance frequencies of both radial and non-radial modes in a spherical resonator have been examined in detail by Mehl [7, 8] and the main results of his work are summarized in this section.

It was found convenient to express the shape perturbation in terms of a series of spherical harmonics such that the new surface is given by

$$r/a = 1 - \varepsilon \sum_{r=0}^{\infty} \sum_{s=-r}^{r} C_{rs} Y_{rs}^{l}(\theta, \xi) \quad (3.7.26)$$

where Y'_{rs} is a normalized spherical harmonic proportional to $P_r^{|s|}(\cos\theta)\exp(\mathrm{i}\xi)$. The most important result of the first-order perturbation theory is that the frequencies of the radial modes are unaffected to that order by any smooth distortion that leaves the volume unchanged. The same is true for the average resonance frequency of the $(2l+1)$ components of each non-radial multiplet but, like the azimuthal modes of the cylinder, the individual components do exhibit first-order shifts. For axisymmetric shape perturbations, the degeneracy is generally reduced to $(l+1)$-fold, while non-axisymmetric perturbations may completely lift the degeneracy. The first-order frequency shifts were found to be sensitive only to coefficients C_{rs} with even indices in the range $2 \leqslant r \leqslant 2l$. For the $l=1$ modes, C_{20}, C_{21} and C_{22} can contribute but C_{22} is unlikely to be important (unless there is a second-harmonic wobble of the machine spindle during manufacture) and, neglecting it, the frequency shifts were found to be

$$\Delta f_{1nm}/f_{1n} = [\varepsilon C_{20}/(20\pi)^{1/2}] \left(\frac{\nu_{1n}^2 + 1}{\nu_{1n}^2 - 2}\right) \lambda'_m \qquad (3.7.27)$$

where λ'_m are characteristic values given by

$$\lambda'_{\pm 1} = (1/2) \pm (3/2)[1 + (8/3)(C_{21}/C_{20})]^{1/2} \qquad (3.7.28)$$
$$\lambda'_0 = -1. \qquad (3.7.29)$$

For the even simpler case of axisymmetric distortions, for which $C_{21} = 0$ also, these reduce to $\lambda'_1 = 2$ and $\lambda'_0 = \lambda'_{-1} = -1$.

When the theory is applied for the radial modes correct to second order in ε, non-zero frequency shifts are found. The magnitude of such terms may be illustrated by the specific case of a spheroidal distortion of the form $r = a(1 - \varepsilon \cos^2\theta)$ which, when the effects of the change in volume are excluded, leads to a perturbation given by

$$\Delta f_{0n}/f_{0n} = (4\nu_{0n}/135)\varepsilon^2 + \mathrm{O}(\varepsilon^3). \qquad (3.7.30)$$

Since the wave equation is separable in spheroidal coordinates (a table of separable coordinates is given in [12]), equation (3.7.30) may also be obtained by expanding an exact solution (M Greenspan, unpublished calculations quoted by Mehl and Moldover [13]; note that the definition of ε there is different from the one used later by Mehl [7] and followed here). The second-order frequency shifts arising from other likely distortions are of similar magnitude and can be found in [7].

3.8 The High-frequency Interferometer

The emphasis of this chapter has been directed towards a normal-mode description of cavity acoustics. The source has been considered only through the mathematical device of an infinitesimal point driver which we were free to place anywhere within the cavity. However, in the approach to high frequencies, where the mean separation of modes in frequency space becomes comparable to the mean linewidth, this theory starts to become cumbersome; many normal modes contribute to the measured signal and convergence of the normal-mode expansion of the Green function is then poor. In practice, devices that are operated at high frequencies are driven by distributed sources that couple efficiently only to normal modes of a particular symmetry, thereby simplifying the problem by eliminating many unwanted modes. By far the most important class of such instruments is the cylindrical interferometer in which the drive transducer forms one complete end; this is the only type of device that shall be considered here.

3.8.1   *Constant or variable pathlength?*

Cylindrical interferometers are categorized as either constant pathlength, and hence variable frequency, or variable pathlength with frequency fixed. The advantages of one kind over the other for particular purposes will be expanded in more detail in Chapter 6. Here it is only necessary to indicate that both possibilities exist and that our theory should be applicable to either. In fact, while both methods have found favour at audio frequencies, the vast majority of high-frequency measurements (say, 1 MHz or above) have used variable path devices with constant excitation frequency. An important factor contributing to this choice is the availability of suitable transducers with which to drive the interferometer, resonant piezoelectric crystals offering only discrete operating frequencies being the most favorable choice at high frequencies. The idealized picture of transducer operation, that one must seek to emulate at high frequencies where many modes could propagate in the interferometer, is one in which the entire end plate executes piston-like motion with velocity $v_t \exp(\mathrm{i}\omega t)$; only longitudinal modes can be excited by such motion. The actual situation when one end of the interferometer is closed by an x-cut quartz crystal vibrating at resonance in its thickness mode may not be too much different from the idealized case.

3.8.2   *Standing versus travelling waves*

Having restricted attention to motion along the axis of the interferometer, one is led to a description of the driven system in terms of a superposition of positive- and negative-going plane waves; indeed such a description will be developed below. However, it is interesting first to demonstrate the mathematical connection between this and the normal-mode description that we have used previously. To do that, we consider the only slightly idealized model of the high-frequency interferometer in which bulk absorption of sound is the dominant loss mechanism and surface losses are neglected. If the source, executing the simple plane motion assumed above, occupies the entire end face at $z = 0$ then the normal-mode expansion of the Green function (equation (3.3.8)) may be integrated over the area of the source to yield the series

$$p_\omega(z) = -i\rho u v_t (kL/\pi^2) \sum_{l=0}^{\infty} \frac{\cos(l\pi z/L)}{\Lambda_l[(kL/\pi)^2 - l^2]} \tag{3.8.1}$$

which may be summed analytically using the methods described by Morse and Feshbach [9] to give

$$p_\omega(z) = -i\rho u v_t \cos[k(z - L)]/\sin(kL). \tag{3.8.2}$$

This is exactly the result that one would obtain by finding the superposition of two plane travelling waves that satisfies the boundary conditions $v_z = v_t$ at $z = 0$ and $v_z = 0$ at $z = L$.

It is worth while examining the behaviour of the system in various regimes of frequency and pathlength. For small values of kL, the interferometer has sharp resonances of Lorentzian lineshape whenever $\mathrm{Re}(kL) = n\pi$ ($n = 1, 2, 3, \cdots$), with the response near to the nth resonance dominated by the normal mode with $l = n$. The acoustic pressure, shown in figure 3.7(*a*) is then a nearly periodic function of z between the end faces, as expected from the factor $\cos[k(z - L)]$ when $-\mathrm{Im}(kL) = \alpha L \ll 1$. However, as the frequency is increased the effect of the absorption coefficient increases ($\alpha \sim f^2$), the resonances broaden and the amplitude of the acoustic pressure, while still showing cyclic behaviour, decays significantly along the length of the cavity (figure 3.7(*b*)). Ultimately at high frequencies and pathlengths long compared with the wavelength, the magnitude of αL can be such that standing waves are hardly formed at all. Behaviour of this kind is shown in figure 3.7(*c*). The acoustic pressure at $z = L$ becomes $2\rho u v_t \exp(-ikL)$, just twice the signal that would be detected in a semi-infinite tube driven at $z = 0$. The 'pressure doubling' near the end face is the usual effect caused by constructive interference between incident and reflected waves. However, the attenuation is such that the waves reflected from the end quickly decay to a level where they are insignificant compared with those that have travelled directly from the source. Thus interference effects are found, in the example illustrated, only within a few wavelengths of the end face. A variable path interferometer driven at a fixed frequency serves as a good illustration. Figure 3.7(*d*) shows the magnitude of the acoustic pressure at the end remote from the source as the pathlength is varied at constant frequency (k fixed). For short paths (αL small), the received signal shows the pronounced resonances expected when the standing wave ratio is high, while at long paths (αL large) the resonances are no longer discernible above the signal travelling directly from the source, and the detected signal varies with amplitude like $\exp(-\alpha L)$ and phase like $2\pi L/\lambda$. All of this information is contained in the normal-mode expansion of the appropriate Green function; however, it is held therein in an inconvenient form. The simple expression obtained from a consideration of interfering waves (or from analytic evaluation of the summation over the normal modes) provides a much more compact description appropriate to regimes of either high or low standing wave ratio.[7] Both these limits provide opportunities for measurement of the speed and attenuation of sound. In the former case one must measure resonance frequencies and linewidths, varying either frequency or pathlength over a narrow range, while in the latter case one must measure amplitude and phase over an extended range of pathlength.

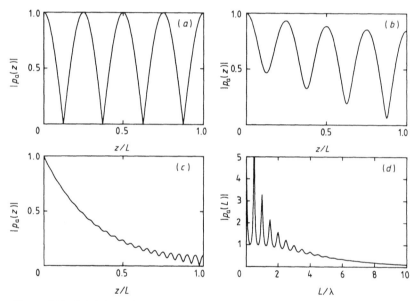

Figure 3.7 Modulus of the acoustic pressure in a cylindrical interferometer: (*a*) as a function of z with $\alpha\lambda = 0$ and $L = 2\lambda$; (*b*) as a function of z with $\alpha\lambda = 0.3$ and $L = 2\lambda$; (*c*) as a function of z with $\alpha\lambda = 0.3$ and $L = 10\lambda$; (*d*) as a function of L/λ with $\alpha\lambda = 0.3$ and $z = L$.

3.8.3 Transmission and reflection losses

In order to account fully for loss mechanisms in the interferometer in a manner that leads to a result more useful than the normal-mode description, we resort to a treatment in terms of a superposition of plane travelling waves. In such a description reflection coefficients are used to describe the change in amplitude and phase upon reflection at either end of the interferometer. Although the meaning of the reflection coefficient χ_r that we will assign to the reflector is straightforward, the meaning of its analogue χ_t at the surface of the transducer requires a little explanation.

In the following, plane piston-like motion of the transducer will be assumed and its velocity $v_t \exp(i\omega t)$ will be a given quantity. Then, if we neglect the thermal boundary layer at the surface of the transducer, an outgoing pressure wave $A \exp(-ik_{01}z)$ will be generated, where $A = (\omega\rho/k_{01})v_t$ and k_{01} is the propagation constant for the plane-wave transmission mode in the cylinder.[8] Since (neglecting the boundary layer) this wave automatically satisfies the boundary condition on the fluid velocity at the surface $z = 0$, any other wave impinging upon that surface must be considered as being reflected without change in amplitude or phase. This ensures that it makes no net contribution to the velocity there. Consequently, when v_t is a given quantity and boundary layers are unimportant, χ_t may be taken as unity.

Actually though, the conventional boundary conditions require the acoustic temperature to vanish at the surface $z = 0$. Consequently, the reflection coefficient χ_t will differ slightly from unity and, in terms of the specific acoustic admittance y_h of the thermal boundary layer, may be written

$$\chi_t = (1 - y_h)/(1 + y_h) \approx (1 - 2y_h). \tag{3.8.3}$$

This ensures that incoming waves, and the reflected propagational and thermal waves to which they give rise, still make no net contribution to the fluid flow at the transducer. Of course, such incoming waves will change the value of the acoustic pressure at the surface of the transducer which must then be driven harder or less hard (depending upon the phase) in compensation, thereby maintaining the given velocity.

The outgoing waves generated directly by the transducer will also be modified slightly by the boundary layer. Both a pressure and a thermal wave, with respective fluid velocities $(1 - y_h)v_t$ and $y_h v_t$ at $z = 0$, will now be generated. Since $(1 - y_h) \approx \chi_t^{1/2}$, the amplitude A of the outgoing pressure wave will be $(\omega\rho/k_{01})v_t\chi_t^{1/2}$ to a good approximation.

Having established the form of the outgoing wave, and defined the reflection coefficients, it is now a simple matter to add up the positive- and negative-going waves in the cylinder. Since the outgoing pressure wave is $A \exp(i\omega t - ik_{01}z)$, the acoustic pressure of the wave returning from the reflector after suffering a single reflection will be $A\chi_r \exp[i\omega t - ik_{01}(2L - z)]$. The total acoustic pressure at any point inside the cylinder will be the sum of these two waves and all their reflections. Dropping the time-dependent factor, we therefore have

$$p(z) = (\omega\rho/k_{01})v_t\chi_t^{1/2}\{\exp(-ik_{01}z) + \chi_r \exp[-ik_{01}(2L-z)]\}$$

$$\times \sum_{n=0}^{\infty} \chi_t^n\chi_r^n \exp(-2ink_{01}L)$$

$$= (\omega\rho/k_{01})v_t\chi_t^{1/2}\left(\frac{\exp(-ik_{01}z) + \chi_r \exp[-ik_{01}(2L-z)]}{1 - \chi_t\chi_r \exp(-2ik_{01}L)}\right) \qquad (3.8.4)$$

which reduces to equation (3.8.2) in the limit $\chi_r, \chi_t \to 1$. Reflection losses are therefore accounted for in a simple manner using reflection coefficients, and losses at the side wall of the interferometer are accounted for through the value of the plane-wave propagation constant. If the walls were rigid and boundary layer losses negligible then k_{01} would take the value of free space: $k = (\omega/u) - i\alpha$. In general k_{01} has a real part equal to ω/u_{01} and an imaginary part equal to $-\alpha_{01}$, where u_{01} and α_{01} are the phase speed and absorption per unit length for the plane-wave transmission mode within the cylindrical waveguide formed by the wall of the interferometer.[9] Thus, when the walls are rigid and the only loss mechanisms other than absorption in the bulk of the fluid are the viscothermal losses at the wall, k_{01} may be obtained from the Kirchhoff–Helmholtz propagation constant (given by equation (2.4.35)):

$$k_{01} = (\omega/u) + (1 - i)\alpha_{KH} - i\alpha. \qquad (3.8.5)$$

The boundary layer terms in equation (3.8.5), and also in the reflection coefficients, will be of diminishing importance at increasing operating frequencies because α_{KH} and y_h both increase only as $f^{1/2}$ while the bulk absorption coefficient α increases like f^2 in non-relaxing gases.

3.8.4 Higher-order transmission modes

The model in which only plane waves propagate within the cavity is an idealized one and many practical arrangements will fail to conform exactly with this simple picture. We must therefore address the issue of what effect the other normal modes of the cavity have when they are excited along with the plane-wave modes. In keeping with the present treatment of the problem in terms of travelling waves we are interested in the generalized transmission modes of the semi-infinite cylindrical waveguide driven by a transducer placed at $z = 0$. These may be derived by separation of the wave equation subject to boundary conditions at the side wall. Assuming idealized homogeneous boundary conditions one finds wavefunctions of the form

$$\phi_{mn}(r, \theta, z) = J_m(\chi_{mn}r/b)[A_{mn} \cos(m\theta) + B_{mn} \sin(m\theta)]\cos(k_{mn}z) \qquad (3.8.6)$$

where χ_{mn} are the turning values of the cylindrical Bessel function of order m (tabulated in §3.6), and k_{mn} is the propagation constant for the m, n mode given by

$$k_{mn} = [k_{01}^2 - (\chi_{mn}/b)^2]^{1/2}. \qquad (3.8.7)$$

Typical behaviour of the real and imaginary parts of k_{mn} is illustrated as a function of $\omega b/u$ in figure 3.8. It can be seen that each of the higher modes has a unique phase speed $u_{mn} = \omega/\text{Re}(k_{mn})$, which is greater than u_{01}, and a cut-off frequency $f_{mn} = (\chi_{mn}/2\pi b)u_{01}$ below which transmission is severely attenuated. In fact this cut-off frequency is identical with the resonance frequency of the $l = 0$ normal mode with the given azimuthal and radial indices, and the divergence of α_{mn} to infinity as f drops towards f_{mn} corresponds with the vanishing of energy transmission along the axis when the wavevector becomes orthogonal with that direction. The plane-wave mode is of course propagational at all frequencies.

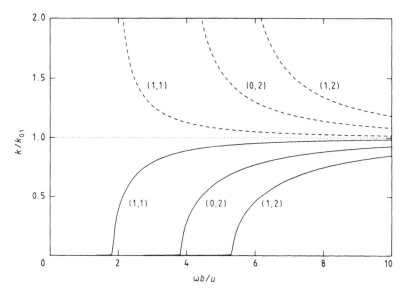

Figure 3.8 Phase speeds and absorption coefficients for selected higher modes in a cylindrical waveguide neglecting surface absorption: ———, $\text{Re}(k_{mn})/\text{Re}(k_{01}) = u_{01}/u_{mn}$; — — — —, $\text{Im}(k_{mn})/\text{Im}(k_{01}) = \alpha_{mn}/\alpha_{01}$. The modes are labelled (m, n).

The constants A_{mn} and B_{mn} in equation (3.8.6) are real quantities giving the amplitude of each mode. They will be determined by the boundary conditions at the face of the transducer and, in the ideal case of piston-like motion of the whole surface, only A_{01} will be non-zero. However, if the active surface of the transducer does not cover the whole surface at $z = 0$, or if its motion is other than that of the ideal piston, then higher modes will be excited. Since these modes have phase speeds and attenuations different from the plane-wave mode, systematic errors can be introduced into measurements of those quantities unless the wave field is properly characterized. One way in which this problem may be avoided is to operate below the lowest cut-off frequency $f_{11} \approx 1.84(u/2\pi b)$, where higher modes are severely attenuated, or at least in the low-frequency regime where the purely longitudinal modes are resolved from all others. Alternatively, when one is interested in high frequencies, where this condition cannot be fulfilled except in tubes of impracticably small diameter, one may seek to use a relatively wide tube in which the lower-order transmission modes all have phase speeds and attenuations close to those of the plane-wave mode. Most practical arrangements can be made to mitigate strongly against any mode with $m \neq 0$ and against modes with large values of n. Thus the high-frequency interferometer will measure an average (or group) speed of a set of unresolved transmission modes dominated by terms with small values of n and, with a suitable choice of transducer geometry and tube diameter, that group speed will closely approximate the phase speed of the plane-wave mode. Calculations for ultrasonic interferometers with rigid walls driven by non-ideal model transducers indicate that errors in the sound speed are unlikely to exceed 0.01 per cent even under conditions where the diameter of the transducer is as small as one-quarter of the tube diameter [10, 11]. In all cases, the errors fall off rapidly as the diameter of the tube is increased.

References

[1] Morse P M and Ingard K U 1968 *Theoretical Acoustics* (New York: McGraw-Hill) pp 556, 585–8
[2] Morse P M and Feshbach H 1953 *Methods of Theoretical Physics* (New York: McGraw-Hill) part II, p 1052–5
[3] Olver F W J (ed.) 1960 *Royal Society Mathematical Tables Volume 7: Bessel Functions part III, zeros and associated values* (Cambridge: Cambridge University Press)
[4] Abramowitz M and Stegun I A (eds) 1964 *Handbook of Mathematical Functions* (Washington, DC: US Government Printing Office)
[5] Mehl J B 1985 *J. Acoust. Soc. Am.* **78** 782
[6] Moldover, M R, Mehl J B and Greenspan M 1986 *J. Acoust. Soc. Am.* **79** 253
[7] Mehl J B 1982 *J. Acoust. Soc. Am.* **71** 1109
[8] Mehl J B 1986 *J. Acoust. Soc. Am.* **79** 278
[9] Morse P M and Feshbach H 1953 *Methods of Theoretical Physics* (New York: McGraw-Hill) part I, pp 413–14
[10] Del Grosso V A 1971 *Acustica* **24** 299

[11] Colclough A R 1973 *Metrologia* **9** 75

[12] Morse P M and Feshbach H 1953 *Methods of Theoretical Physics* (New York: McGraw-Hill) part I, pp 655–64

[13] Mehl J B and Moldover M R 1981 *J. Chem. Phys.* **74** 4062

[14] Weston D E 1953 *Proc. Phys. Soc.* B **66** 695

[15] Shields F D, Lee K P and Wiley W J 1965 *J. Acoust. Soc. Am.* **37** 724

General references

Morse P M and Ingard K U 1968 *Theoretical Acoustics* (New York: McGraw-Hill) pp 554–76

Moldover M R, Mehl J B and Greenspan M 1986 *J. Acoust. Soc. Am.* **79** 253

Notes

1. $\Phi^*(r)$ denotes the complex conjugate of $\Phi(r)$.
2. This applies especially to the case of a source with constant surface displacement amplitude in the frequency range of the measurements ($S_\omega \propto \omega$); strictly, the resonance term should be modified for other source characteristics.
3. For example, the wave equation is not separable in toroidal coordinates.
4. To avoid confusion with the coordinate y, the specific acoustic admittance will be denoted by β in this subsection.
5. That is, for metal tubes with wall thickness greater than $b/10$.
6. This is a consequence of the numbering scheme by which the zero-frequency mode is denoted by $n = 1$.
7. The standing wave ratio is defined as the ratio of the acoustic pressure amplitude at an antinode to that at an adjacent node.
8. The factor $(1 + i\omega\tau_V)$, of equation (2.3.36), is taken as unity here. However, $\omega\rho/k_{01}$ is not set equal to ρu in recognition of the fact that the imaginary part of k_{01} may be inflated above order $\omega^2\tau_V/u$ by the boundary layer losses at the side walls of the cylinder.
9. The phase speed is defined as the speed at which points of constant phase propagate in a given direction—in this case along the axis of the cylinder.

4

Relaxation Phenomena

4.1 Introduction

In formulating the equations describing the propagation of sound in an
ideal fluid we assumed that local thermodynamic equilibrium was achieved
instantaneously in the fluid. The local temperature and pressure were
assumed to follow the local density along an isentropic path. However, we
have already seen that diffusion of heat and momentum causes the thermodynamic path of the acoustic cycle to deviate from an isentropic one.
Often, there are other relaxation mechanisms operating in the fluid which
give rise to a delay between the imposition of a density change and the
establishment of the equilibrium temperature and pressure. We will show
that if the relaxation times characterizing these mechanisms are short compared with $1/\omega$ then the internal friction of bulk viscosity results with η_b
real and independent of frequency. When this term is included, the hydrodynamic equations can be applied with confidence. However, if the delay
in the attainment of local equilibrium is of the order of $1/\omega$ then large
effects can result that are not adequately described by the theory so far
derived.

We first compare the predictions of the Navier–Stokes equations with
other theoretical predictions and with the experimental results for monatomic gases at frequencies approaching those of molecular collisions. We
will find that the onset of translational dispersion marks the end of the
regime in which the hydrodynamic theory can be rendered accurate. We
also mention the existence of additional sound absorption in fluid mixtures.
Next, the important relaxation effects associated with energy transfer
between translational and internal degrees of freedom will be considered.
A theoretical description is greatly simplified if translational, rotational,
and vibrational relaxation processes are treated independently. The vibrational modes of molecules generally have relaxation times much greater
than the mean time between collisions and comparatively simple theory is
therefore applicable. However, rotational relaxation is usually sufficiently

fast for the domain of its importance to overlap that of translational relaxation; some of the theoretical difficulties that arise in this case are considered in the latter part of §4.3. Finally, we will consider the effects of chemical and structural rearrangements in the fluid.

4.2   Translational Relaxation

In this section we will consider only monatomic fluids and thereby temporarily avoid the complications arising from internal degrees of freedom; in addition, we shall restrict attention to propagational (sound-like) waves. In this case, the experimental results for the speed and absorption of sound provide a test of our treatment of translational energy transfer between molecules essentially unaffected by other factors.

4.2.1   Theoretical predictions

The hydrodynamic equations introduced in Chapter 2, and referred to as Navier–Stokes, are those that arise when the fluxes of heat and momentum are linearized in the spatial gradients of temperature and velocity. For simple-harmonic sound we derived an exact solution, equation (2.3.31), the properties of which were examined at low frequencies. In this context, 'low frequency' refers to the regime in which $\omega\tau_h$ and $\omega\tau_s$ are small compared with unity. Here, $\tau_h = D_h/u_0^2$ and $\tau_s = D_s/u_0^2$ are relaxation times that characterize the transfer of heat and momentum. In that regime, the speed of sound is independent of frequency and the absorption per wavelength, $\mu = \alpha\lambda$, is linear in ω. At higher frequencies, equation (2.3.31) predicts that the phase speed u should be greater, and μ/ω less, than the corresponding low-frequency values. For an ideal monatomic gas ($\gamma = 5/3$) the coefficients of viscosity and thermal conductivity are, to a good approximation, related by $\varkappa/\eta c_V = 5/2$, and hence $\tau_h = 3\tau_s/2$. Actually, it is convenient to eliminate both quantities in favour of the characteristic time

$$\tau_c = 4\eta/5p. \qquad (4.2.1)$$

Since, for a hard-sphere gas, τ_c is equal to the mean time between collisions, we shall take $f_c = \tau_c^{-1}$ as an operational definition of the molecular collision frequency in both real and model gases.

The Navier–Stokes theory applies to a continuum fluid and therefore only at frequencies small compared with those of molecular collisions. In gases, a regime of high f/p is experimentally accessible in which f/f_c is not small (i.e. where the wavelength is comparable with the mean free path) and where the continuum theory cannot be expected to hold; indeed we shall see that the experimental results do conflict with that theory. There have been several attempts to improve the description of sound propagation in rarefied gases by returning to the fundamental equation of kinetic theory—that of Boltzmann. For the sake of brevity, we shall concentrate just on the results of the three main approaches that have been adopted, omitting the mathematical details of each.

The first approach involves a strategy for solving the (linearized) Boltzmann equation, known as the Chapman–Enskog method [1], which takes the form of successive approximations. The first of these approximations yields Euler's equation and the second the Navier–Stokes equation and Fourier's equation, while the third and fourth approximations are known as the Burnett and super-Burnett equations [1–3]. Each of these different equations may be taken as a starting point from which expressions for the speed and absorption of sound can be obtained; in Chapter 2 we examined the first- and second-order cases. The higher approximations involve expressions for the stress and heat flux which include powers of the corresponding spatial gradients greater than the first. They lead to new (and much more complicated) expressions for u and α. Unfortunately, the results of increasing order do not appear to converge at frequencies approaching f_c.

A different method was followed by Wang Chang and Uhlenbeck who solved the linearized Boltzmann equation *directly* for forced simple-harmonic plane waves (without first obtaining general expressions for the stress and heat flux). This method yielded an expansion of the propagation constant in powers of f/f_c which also shows poor convergence at very high frequencies. Pekeris *et al* [5] obtained improved results by resorting to numerical methods but even these did not seem to converge at frequencies above about 20 per cent of the molecular collision frequency.

Despite the difficulties, these two approaches may be useful in the regime where u/u_0 is just starting to rise above unity. Greenspan [4] gives series expansions for the reduced propagation constant $\Gamma = ku_0/\omega$ that may be written in powers of $\omega\tau_c$ as follows:

Navier–Stokes $\quad \Gamma = 1 - \frac{7}{8}\mathrm{i}(\omega\tau_c) - \frac{141}{128}(\omega\tau_c)^2 + 1.522\mathrm{i}(\omega\tau_c)^3 - \cdots$ (4.2.2)

Burnett $\quad \Gamma = 1 - \frac{7}{8}\mathrm{i}(\omega\tau_c) - \frac{215}{128}(\omega\tau_c)^2 + 4.104\mathrm{i}(\omega\tau_c)^3 - \cdots$ (4.2.3)

Super-Burnett $\quad \Gamma = 1 - \frac{7}{8}\mathrm{i}(\omega\tau_c) - \frac{215}{128}(\omega\tau_c)^2 + 5.609\mathrm{i}(\omega\tau_c)^3 - \cdots$ (4.2.4)

Wang Chang and $\quad \Gamma = 1 - \frac{7}{8}\mathrm{i}(\omega\tau_c) - \frac{215}{128}(\omega\tau_c)^2 + 5.034\mathrm{i}(\omega\tau_c)^3 - \cdots$. (4.2.5)
Uhlenbeck

All of these results predict the same low-frequency values of the speed and absorption of sound. All but the Navier–Stokes agree up to order $(\omega\tau_c)^2$, but none of the solutions agree exactly to higher order. These observations suggest that the prediction of the Navier–Stokes theory with regard to

Re(Γ), which gives the phase speed, fails in leading order but that the higher approximations are accurate as far as that term.

The third approach which has been followed involves the use of mathematical models of the Boltzmann equation [6], from which essentially exact solutions for plane simple-harmonic sound waves can be computed, in place of the approximate solutions of the full Boltzmann equation considered above. These so-called kinetic models yield results that appear to be valid at frequencies up to and above f_c without the use of a great many terms (or moments) in the model equation.

4.2.2 Comparison with experiment

The most comprehensive sets of experimental measurements that might discriminate amongst the various theories for the propagational wave in a monatomic gas are those of Greenspan [7]. These were measured at the single frequency of 11 MHz using pressures between about 0.6 and 100 kPa to vary the ratio f/f_c. Greenspan's results are in good agreement with those of Meyer and Sessler [8] which were obtained in the range $0.01 \leqslant f/f_c \leqslant 10$ using frequencies of order 100 kHz and correspondingly low pressures.

The experimental results for the phase speed and absorption, expressed in terms of the reduced propagation constant Γ, are compared with the theoretical predictions in figure 4.1. In terms of Γ, the absorption is given by $-\mathrm{Im}(\Gamma) = \alpha\lambda_0/2\pi$, where $\lambda_0 = u_0/f$, and the phase speed is given by $\mathrm{Re}(\Gamma) = u_0/u$. On the scale of the plot, the experimental absorption results and all of the theories are in agreement up to about 1 per cent of the molecular collision frequency (in argon at 300 K and 100 kPa, τ_c^{-1} is about 5 GHz). There is not a consensus between the theories at higher frequencies and the experimental results do not strongly favour any one of them below $f = f_c$. However, while some of the other predictions appear better for either phase speed or absorption, the 11-moment kinetic model of Sirovich and Thurber [6] is the only one that offers a reasonable approximation to both quantities. At frequencies above f_c (not shown in figure 4.1), the phase speed and absorption both level off with $u_0/u \approx 0.5$ and $\alpha\lambda_0/2\pi \approx 0.25$ [8]. In that regime, only the kinetic models follow the experimental results.

The predictions of the kinetic models are dependent both upon the number of moments retained and upon the intermolecular forces assumed. The variations observed in going from 8 to 11 moments and from Maxwellian molecules[1] to hard spheres are of the same order as the discrepancies observed between the experimental results and the 11-moment model illustrated in figure 4.1 [6]. Given these variations, the level of agreement shown there for both phase speed and absorption must be considered good.

In liquids the molecular collision frequency is well above the maximum frequencies experimentally accessible. However, experimental measurements of the speed and absorption of sound in monatomic liquids have been reported both at ultrasonic frequencies (~ 1 MHz) [9] and at the ultra-high frequencies (~ 1 GHz) [10, 11] of the Brillouin scattering method. No dispersion has been found across this wide frequency range. For example, the speed of sound in liquid xenon measured by Brillouin scattering at 3.25 GHz [10] agrees with ultrasonic speeds measured at

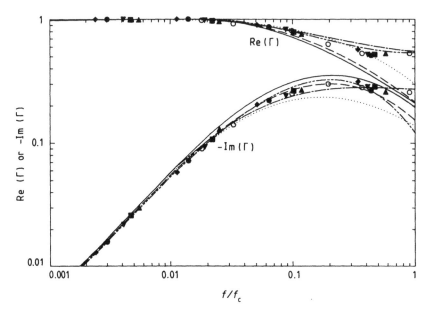

Figure 4.1 Absorption and dispersion of sound in monatomic gases. Experimental results of Greenspan [4]: ▲, He; ▼, Ne; •, Ar; ♦, Kr; ■, Xe. Experimental results of Meyer and Sessler for argon: ○. Theoretical predictions: —————, Navier–Stokes; — — — —, Burnett; ⋯⋯⋯, super-Burnett; — – – —; 483-moment numerical solution for a Maxwellian gas [5]; — — — —, 11-moment kinetic model for a hard-sphere gas [6].

1 MHz [9] to within experimental uncertainty of order 0.2 per cent. These observations are in agreement with the theoretical prediction based on the Navier–Stokes equations. Absorption coefficients obtained from the linewidths of the Brillouin doublet indicate that the bulk viscosity in monatomic liquids is not negligible compared with the ordinary shear viscosity. In liquid xenon, η_b was found to be almost independent of temperature with the ratio $\eta_b/\eta \approx 0.8$ at the triple point [10]. The results for argon and krypton also give ratios η/η_b of order unity.

4.2.3 Mixtures

Thermal conduction and viscous damping are not the only processes that can lead to sound absorption in monatomic mixtures. In the presence of the density and temperature gradients imposed by the sound field, the components of the mixture will tend to undergo partial fractionation but the concentration gradients thereby established will be damped by diffusion. The extra absorption α_{diff} arising from this process has been evaluated by Kohler for the non-dispersive region of a binary mixture of ideal gases (not necessarily monatomic) [23]. For the mixture $(1 - x)A + xB$, his result may be written

$$\alpha_{\text{diff}} = (\omega^2/2u^3)\gamma x(1 - x)D\{[(M_B - M_A)/M] + [(\gamma - 1)/\gamma x(1 - x)]K_T\}$$

$$(4.2.6)$$

where M_A and M_B are the molar masses of the two components, D is the binary diffusion coefficient and K_T is the thermal diffusion ratio. In this regime the sound speed is given by equations (1.2.9) and (1.2.13). Experimental results for u and $(\alpha - \alpha_{\text{cl}})$ in monatomic gases are in good agreement with values calculated from these expressions [24].

For mixtures of components that differ greatly in molar mass (e.g. helium and xenon), the hydrodynamic theory is predicted to break down at frequencies smaller than those that we have been talking about for single-component gases by a factor of order $(M_A/M_B)^{1/2}$ (where A is the heavier component). The possibility of two sound-like modes above some critical frequency has been proposed [25] but firm experimental evidence for this is currently lacking.

4.3 Molecular Thermal Relaxation

4.3.1 Frequency dependence of the heat capacities

The pressure exerted by a fluid is determined by the translational modes of motion of the molecules. If the density is suddenly altered then these degrees of freedom will adjust almost immediately because few collisions are required for the equilibration of translational energy. Accordingly, for the present we neglect the translational effects discussed in the previous section. However, the instantaneous pressure increment δp, just after a small change $\delta\rho$ in density, will still not necessarily be equal to the value

$$\delta p = (1/\varkappa_S)(\delta\rho/\rho) = (\gamma/\varkappa_T)(\delta\rho/\rho) \qquad (4.3.1)$$

determined by the isentropic equation of state because of the finite time required for the exchange of energy between translational and internal modes of molecular motion (here $\varkappa_S$ and $\varkappa_T$ are respectively the isentropic and isothermal compressibility). This situation is illustrated in figure 4.2 where we plot the pressure increment following an instantaneous compression at time $t = 0$ for a fluid with one internal degree of freedom undergoing relaxation. The initial value of δp just after the compression is given by $(\gamma_{\text{eff}}/\rho\varkappa_T)\delta\rho$, where γ_{eff} is an effective heat capacity ratio of the fluid with the relaxing degree of freedom 'frozen' out. In the time following the compression, the pressure decays towards the value of equation (4.3.1) as energy flows into the internal mode at a rate governed by the time constant of the relaxing degree of freedom. We see that the work done in the compression exceeds the stored energy available when the system is allowed to reach equilibrium by an amount $(\gamma_{\text{eff}} - \gamma)(\delta\rho/\rho\varkappa_T)$ per unit volume; although the process was adiabatic it was not reversible.

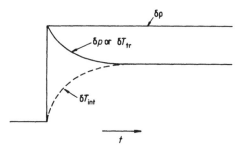

Figure 4.2 Relaxation of pressure, translational temperature and characteristic temperature of internal modes following an instantaneous compression.

We now examine more closely the path along which the system approaches equilibrium. Since the isothermal compressibility is unaffected by the internal degrees of freedom we give our attention to the dynamic value of the heat capacity ratio. It is useful to introduce the concept of translational (T_{tr}) and internal (T_{int}) temperatures such that an infinitesimal change in the energy at constant volume may be written as the sum of two terms, one for the energy of translation and one for that of the internal mode:

$$dE = (C_V - C_{\text{int}}) \, dT_{\text{tr}} + C_{\text{int}} \, dT_{\text{int}}. \tag{4.3.2}$$

Here, C_V is the total (static) constant-volume heat capacity and C_{int} is that part of the heat capacity arising from the storage of energy by the internal mode. For slow changes, the translational and internal temperatures remain in equilibrium with each other by exchanging energy. However, this exchange takes time; rapid changes in T_{tr} cannot be followed faithfully by T_{int}.

We assume that, following a small displacement $\delta E_{\text{int}} = E_{\text{int}}(T_{\text{int}}) - E_{\text{int}}(T_{\text{tr}})$,[2] the energy held in the internal mode approaches equilibrium in accordance with the first-order equation

$$(dE_{\text{int}}/dt) = -\delta E_{\text{int}}/\tau \tag{4.3.3}$$

where τ is the relaxation time constant. If the displacement is small then the heat capacities may be taken as constant and equation (4.3.3) is equivalent to

$$(dT_{\text{int}}/dt) = -(T_{\text{int}} - T_{\text{tr}})/\tau. \tag{4.3.4}$$

If the internal temperature is displaced from that of translation and T_{tr} is maintained constant (by contact with a heat bath) then T_{int} relaxes along a constant-volume constant-pressure path with time constant τ. However, following the compression illustrated in figure 4.2, the internal and translational temperatures differ, and *both* relax towards equality along an adiabatic path at constant volume. Under those conditions we may write

$$(C_V - C_{\text{int}})T_{\text{tr}} + C_{\text{int}}T_{\text{int}} = E_0 \tag{4.3.5}$$

where E_0 is constant, and use this expression to eliminate the translational temperature from equation (4.3.4):

$$(dT_{\text{int}}/dt) = -[T_{\text{int}} - (E_0/C_V)]/\tau'. \tag{4.3.6}$$

Here τ' is a new relaxation time given by

$$\tau' = [(C_V - C_{\text{int}})/C_V]\tau \tag{4.3.7}$$

and we see that the observed relaxation time depends upon the conditions. If instead of maintaining constant volume after the initial compression, adiabatic constant-pressure conditions prevailed then the internal and translational temperatures would approach each other with an apparent relaxation time given by

$$\tau'' = [(C_p - C_{\text{int}})/C_p]\tau. \tag{4.3.8}$$

In the presence of sound, the density is not subjected to instantaneous change but varies periodically in time. The question now arises as to which apparent relaxation time is appropriate. In our present approximation (i.e. neglecting diffusion of heat and momentum), the acoustic cycle is certainly adiabatic, but density, temperature and pressure all vary. Fortunately, in the case of simple-harmonic sound, all of the acoustic variables fluctuate as $\exp(i\omega t)$ so that $T_{\text{int}} = T + T_{a,\text{int}}\exp(i\omega t)$ and $T_{\text{tr}} = T + T_{a,\text{tr}}\exp(i\omega t)$, where T is the equilibrium temperature. Substituting in equation (4.3.4) we then obtain

$$(T_{\text{int}} - T) = (T_{\text{tr}} - T)/(1 + i\omega\tau) \tag{4.3.9}$$

from which it follows that

$$(dT_{int}/dT_{tr}) = (1 + i\omega\tau)^{-1}. \tag{4.3.10}$$

We next define an effective constant-volume heat capacity $C_V(\omega)$ for angular frequency ω by

$$dE = C_V(\omega)\, dT_{tr}. \tag{4.3.11}$$

Comparing this with equation (4.3.2) we have

$$dE = [(C_V - C_{int}) + C_{int}(dT_{int}/dT_{tr})]\, dT_{tr} = C_V(\omega)\, dT_{tr} \tag{4.3.12}$$

from which the derivative may be eliminated with (4.3.10) to yield

$$C_V(\omega) = (C_V - C_{int}) + C_{int}/(1 + i\omega\tau) = C_V - i\omega\tau C_{int}/(1 + i\omega\tau). \tag{4.3.13}$$

Thus, when a single mode undergoes relaxation, the ratio $C_V(\omega)/C_V$ is given in terms of the isothermal relaxation time τ in a manner analogous to the transfer function of a single-pole low-pass filter in AC circuit theory.

Since the difference between the constant-pressure and constant-volume heat capacities is independent of internal degrees of freedom, C_p is given by an expression having exactly the same form as equation (4.3.13). The ratio $\gamma_\omega = C_p(\omega)/C_V(\omega)$ of the effective heat capacities is therefore given by

$$\frac{\gamma_\omega}{\gamma} = \frac{1 - [i\omega\tau\Delta/(1 + i\omega\tau)]}{1 - \gamma[i\omega\tau\Delta/(1 + i\omega\tau)]}$$

$$= \frac{[1 + \omega^2\tau^2(1 - \Delta)(1 - \gamma\Delta)] + i\omega\tau(\gamma - 1)\Delta}{[1 + \omega^2\tau^2(1 - \gamma\Delta)^2]} \tag{4.3.14}$$

where $\Delta = C_{int}/C_p$. The real and imaginary parts of γ_ω/γ are plotted in figure 4.3 against the logarithm of $\omega\tau$. We see that $\text{Re}(\gamma_\omega/\gamma)$ increases from unity at low frequencies, initially as $(\omega\tau)^2$, passes through a point of inflexion, and approaches an upper limit equal to $(C_p - C_{int})/(C_V - C_{int})$. In this limit the internal mode has ceased to participate in the acoustic cycle. The angular frequency ω_{inf} corresponding to the point of inflexion on the curve satisfies $\omega_{int}\tau(1 - \gamma\Delta) = \omega_{inf}\tau' = 1$. The imaginary part of γ_ω/γ increases, initially as $\omega\tau$, from zero as $\omega = 0$ but reaches a maximum at $\omega = \omega_{inf}$. At still higher frequencies $\text{Im}(\gamma_\omega/\gamma)$ declines again towards zero, ultimately as $(\omega\tau)^{-1}$. The phase angle ϕ in the complex plane given by

$$\tan\phi = \frac{\omega\tau(\gamma - 1)\Delta}{1 + \omega^2\tau^2(1 - \Delta)(1 - \gamma\Delta)} \tag{4.3.15}$$

is a maximum at angular frequency $\omega_{max} = \omega_{inf}[(1 - \gamma\Delta)/(1 - \Delta)]^{1/2}$. Tan φ is, to a good approximation, equal to the fraction of the acoustic energy dissipated in the the relaxing mode per unit advance in the phase; its maximum value can be as large as about 0.15. The medium is now a dispersive one. We shall see that the speed of sound is determined mainly by the real part of γ_ω so that the phase speed increases from u_0 at zero frequency, rapidly for angular frequencies near to ω_{inf}, and approaches an upper limit asymptotically as $\omega \to \infty$. The attenuation of the sound wave per cycle is substantial in the dispersive region, but drops rapidly on either side of ω_{max}, approaching zero at the extremes.

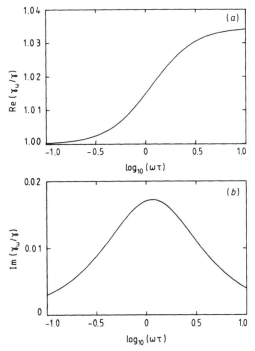

Figure 4.3 γ_ω/γ for a system with one internal mode undergoing relaxation ($\gamma = 1.3$, $\Delta = 0.1$): (a) $\mathrm{Re}(\gamma_\omega/\gamma)$; ($b$) $\mathrm{Im}(\gamma_\omega/\gamma)$.

Since, on a macroscopic scale, the mean flow of energy to and from the internal modes of the molecules varies during the acoustic cycle purely as a result of the fluctuations in temperature, the relaxation process is known as thermal relaxation. On a molecular scale, the exchange of energy between internal and translational modes can take place only by means of collisions between molecules. For gases, where the time τ_c between binary collisions varies rapidly with pressure ($\tau_c \sim 1/p$), the relaxation efficiency is often expressed in terms of the collision number

$$\zeta = \tau/\tau_c \tag{4.3.16}$$

which is nearly independent of pressure. Unfortunately, τ_c has no unique definition for a gas composed of real molecules, and two definitions are in common use. Both are derived from the expression

$$\tau_c = (1/4p\sigma^2)(mkT/\pi)^{1/2} \tag{4.3.17}$$

which applies to a gas composed of hard spheres of diameter σ. In practice either a temperature-independent estimate of σ is employed or the molecular diameter is eliminated in favour of the experimental viscosity using the hard-sphere expression $\eta = (5/16\sigma^2)(mkT/\pi)^{1/2}$. In the latter case, equation (4.3.17) simplifies to the operational definition, $\tau_c = 4\eta/5p$, introduced in §4.2 (equation (4.2.1)). Strictly, the concept of a collision number is meaningful only in gases at low or moderate pressures where multibody collisions may be neglected.

4.3.2 Multiple internal modes

Generally, the thermodynamic energy E may be written as the algebraic sum of contributions stored as translational kinetic energy, intermolecular potential energy, and as potential and kinetic energy in the internal degrees of freedom. For molecules with more than one internal mode undergoing relaxation it is therefore straightforward to extend the definition of an effective heat capacity by generalization of equation (4.3.12):

$$C_V(\omega) = (C_V - C_{\text{int}}) + \sum_n C_n(\mathrm{d}T_n/\mathrm{d}T_{\text{tr}}). \tag{4.3.18}$$

Here, we denote the contribution to the heat capacity arising from the nth mode by C_n, the characteristic temperature of that mode by T_n, and the total static contribution of the internal modes by $C_{\text{int}} = \sum_n C_n$. We must, however, re-examine the manner in which internal energy is exchanged with translations. Two limiting cases can be identified. At one extreme, parallel excitation may occur in which each internal mode communicates independently with translations. At the other extreme, we can imagine that there is highly efficient intramolecular energy exchange between all the internal modes but that just one mode communicates with translations on a significant timescale. This is the model of series excitation. It turns out that neither of these two extremes represents a satisfactory approximation to actual microscopic events. However, a synthesis of the two cases describes quite well the *macroscopic* consequences of molecular energy transfer.

We first consider parallel excitation with all of the degrees of freedom independent. In this case an equation of the form of (4.3.10) may be written for each internal mode. It then follows from (4.3.18) that, in response to a disturbance having a harmonic variation in time, the contribution of the nth internal mode to the effective heat capacity is given by the zero-frequency contribution of the mode, C_n, divided by the factor $(1 + i\omega\tau_n)$. Consequently, the total effective heat capacity at constant volume is given by

$$C_V(\omega) = (C_V - C_{\text{int}}) + \sum_n [C_n/(1 + i\omega\tau_n)]. \qquad (4.3.19)$$

Since the difference $C_p - C_V$ is not influenced by the internal degrees of freedom, $C_p(\omega)$ is given by an expression having the same form as equation (4.3.19) and the effective ratio of heat capacities at frequency $\omega/2\pi$ is given by

$$\frac{\gamma_\omega}{\gamma} = \frac{1 - \sum_n [i\omega\tau_n/(1 + i\omega\tau_n)]\Delta_n}{1 - \gamma\sum_n [i\omega\tau_n/(1 + i\omega\tau_n)]\Delta_n} \qquad (4.3.20)$$

where the unsubscripted γ denotes the value of the ratio under static conditions, and $\Delta_n = C_n/C_p$. Equation (4.3.20) applies to any fluid with parallel excitation of internal modes. The behaviour of γ_ω is now correspondingly more complicated than for the case of a single mode. As the frequency is increased each of the internal modes drops out of the acoustic cycle in turn when ω passes through the inverse of the corresponding relaxation time. If the relaxation times were well separated then $\text{Re}(\gamma_\omega/\gamma)$ would rise smoothly in stages as each mode dropped out, with the ascents being accompanied by peaks in the dissipation.

If only a small fraction of the total heat capacity is relaxing then it is appropriate to take an expansion of equation (4.3.20) in powers of the quantities Δ_n:

$$\gamma_\omega/\gamma = 1 + (\gamma - 1)\sum_n [(i\omega\tau_n + \omega^2\tau_n^2)/(1 + \omega^2\tau_n^2)]\Delta_n + \cdots. \qquad (4.3.21)$$

Here we have resolved the coefficient of Δ_n into real and imaginary parts. In the case of a single internal mode, the radius of convergence, $\Delta < (1/\gamma\omega\tau)(1 + \omega^2\tau^2)^{1/2}$, includes all physically significant values, and if Δ is small then the linear approximation will be accurate at any frequency. Similar arguments apply in the case of multiple relaxing modes. At low frequencies, a direct expansion of equation (4.3.20) in powers of ω (valid for any set of the Δ_n) is more useful:

$$\gamma_\omega/\gamma = 1 + i(\gamma - 1)\sum_n \tau_n\Delta_n\omega$$

$$+ (\gamma - 1)\left[\sum_n \tau_n^2\Delta_n - \gamma\left(\sum_n \tau_n\Delta_n\right)^2\right]\omega^2 + \cdots. \qquad (4.3.22)$$

For a single relaxing mode, the radius of convergence of this series is $\omega\tau < (1 - \gamma\Delta)^{-1}$.

In the case of series excitation, we assume that just one mode, denoted by $n = 1$, undergoes exchange with translations. Usually, this is the mode lowest in energy. It is a simple matter to show that if intramolecular exchange of energy between this mode and all of the other internal degrees of freedom is very rapid then the system exhibits a single overall relaxation time τ related to τ_1 by

$$\tau = (C_{int}/C_1)\tau_1. \qquad (4.3.23)$$

In quantum mechanics, simple series excitation is not possible because the quantum to be surrendered by one mode is generally of unequal size to that required to excite another. Consequently, rearrangement of internal energy should occur efficiently only during collisions where any excess or deficit on the transaction can be absorbed by translational kinetic energy.

Clearly, a complete picture of molecular energy transfer would be extremely complicated. However, acoustic measurements are only capable of revealing a small number of relaxation times, and simplified theory appears to be adequate. It is often useful to consider the effects of rotational degrees of freedom separately from the remainder of the internal modes. There are several reasons for this. First, for molecules other than hydrogen, it turns out that the relaxation collision numbers for the rotational degrees of freedom are small—typically less than 10. Second, again except for hydrogen, the rotational modes are fully excited at accessible temperatures so that the contribution to the molar heat capacity is simply $\frac{1}{2}R$ for each rotational mode. In contrast, the collision numbers of the vibrational modes are often very large and there is usually only partial excitation at the temperatures of interest. Furthermore, the separation of rotational and vibrational motion is a good approximation for purposes of expressing the internal heat capacity in molecular terms. Thus, rotational relaxation is often treated as a set of processes occurring in parallel with other relaxation mechanisms. Usually, it appears that the vibrational modes are strongly coupled with each other on the timescale of thermal relaxation so that a single overall vibrational relaxation time τ_{vib} is observed. Occasionally, two relaxation processes can be discerned amongst the vibrational modes [19]. A satisfactory description for our purposes is therefore afforded by treating thermal relaxation as a small number of parallel processes, each of which represents some group of internal modes that are coupled together as if in series.

4.3.3   The bulk viscosity

The bulk viscosity is a consequence of the irreversible nature of the acoustic cycle. This arises because the instantaneous value of the acoustic pressure p_i differs from, but relaxes towards, the hypothetical value that would be achieved if the fluid were in local equilibrium. To a first-order approximation the effects of the bulk viscosity and those of the classical viscothermal processes enter independently; it will therefore be sufficient to continue our neglect of thermal conduction and viscous damping in the following. Thus, the acoustic pressure p_a of Chapter 2 is identical with the hypothetical value

$$p_a = (\gamma/\rho \varkappa_T)\rho_a \qquad (4.3.24)$$

and the stress tensor is diagonal with elements

$$\mathcal{T}_{ii} = p + p_i = (p + p_a) - \eta_b(\nabla \cdot v) \qquad (4.3.25)$$

where to leading order $\nabla \cdot v = -(\varkappa_T/\gamma)(\partial p_a/\partial t)$. We now show that, at sufficiently low frequencies, the small difference $p_i - p_a$ is indeed given by

$$p_i - p_a = \eta_b(\varkappa_T/\gamma)(\partial p_a/\partial t) \qquad (4.3.26)$$

in accordance with the phenomenological definition of the bulk viscosity.

For simple-harmonic sound the hypothetical rate of compression is, from equation (4.2.24),

$$(\partial p_a/\partial t) = \gamma(i\omega/\rho \varkappa_T)\rho_a = \gamma(i\omega/\rho \varkappa_T)\rho_0 \exp(i\omega t) \qquad (4.3.27)$$

where ρ_0 is the amplitude of the density fluctuations, while the actual rate of compression is

$$(\partial p_i/\partial t) = \gamma_\omega(i\omega/\rho \varkappa_T)\rho_0 \exp(i\omega t). \qquad (4.3.28)$$

Integrating the difference between these two expressions, subject to $(p_i - p_a) \to 0$ as $\omega \to 0$, and eliminating ρ_a with equation (4.3.24) we find

$$p_i - p_a = [(\gamma_\omega/\gamma) - 1]\, p_a = (1/i\omega)[(\gamma_\omega/\gamma) - 1]\, (\partial p_a/\partial t). \qquad (4.3.29)$$

Now, if $\omega\tau_n \ll 1$ for all n then powers of $\omega\tau_n$ above the first may be neglected in equation (4.3.22) so that $[(\gamma_\omega/\gamma) - 1]$ is small and purely imaginary; in this low-frequency regime the amplitude of the acoustic pressure is unaffected in leading order but there is a small phase difference between p_a and ρ_a. Combining equations (4.3.22) and (4.3.29) we obtain

$$p_i - p_a = (\gamma - 1)(\partial p_a/\partial t) \sum_n \tau_n \Delta_n \qquad (4.3.30)$$

and see that, provided $\omega\tau_n \ll 1$ for all n, (4.3.26) is indeed obeyed with a bulk viscosity given by

$$\eta_b = [(\gamma - 1)/\varkappa_S] \sum_n \tau_n \Delta_n = [(\gamma - 1)\rho u^2] \sum_n \tau_n \Delta_n. \qquad (4.3.31)$$

When this is included in the problem, and the criterion $\omega\tau_n \ll 1$ is met, all of the expressions derived in Chapter 2 may be applied with confidence and relaxation of internal modes need be considered no further.

4.3.4   Absorption and dispersion

The bulk viscosity accounts for relaxation absorption accurately only at frequencies such that ω is small compared with the longest relaxation time of the internal modes. In that regime the phase speed is unaffected by relaxation. However, at higher frequencies we must formulate a new description of sound propagation for a relaxing fluid. We continue, for the moment, to neglect the classical viscothermal effects so that we are faced with a fluid that is ideal in every respect except that it has a complex and frequency-dependent heat capacity ratio. The wave equation equivalent to equation (2.2.12) is now

$$[\nabla^2 - (\rho \chi_T / \gamma_{\text{eff}})(\partial^2 / \partial t^2)] \, p_{\text{a}} = 0 \qquad (4.3.32)$$

where γ_{eff} is an effective value of the heat capacity ratio. For the special case in which we are interested, that of simple-harmonic sound, we may identify γ_{eff} as γ_ω and replace ∇^2 and $-k^2$ and $\partial^2 / \partial t^2$ by $-\omega^2$. The solutions, written in terms of the reduced propagation constant $\Gamma = k u_0 / \omega$, are then given by

$$\Gamma^2 = (u_0 / u)^2 [1 - i(\mu / 2\pi)]^2 = \gamma / \gamma_\omega \qquad (4.3.33)$$

where $\mu = \alpha \lambda$ is the absorption per wavelength. Experimental measurements of u and α serve to determine both the real and imaginary parts of γ_ω and can be analysed in terms of parameters τ_n and Δ_n by direct comparison with equation (4.3.20). The phase speed and absorption are given by

$$u / u_0 = 1 / \text{Re}(\Gamma) \qquad (4.3.34)$$

$$\alpha \lambda_0 / 2\pi = -\text{Im}(\Gamma) \qquad (4.3.35)$$

where $\lambda_0 = u_0 / f$, while the absorption per actual unit advance in the phase is

$$\mu / 2\pi = -\text{Im}(\Gamma) / \text{Re}(\Gamma). \qquad (4.3.36)$$

For a single mode it is sometimes useful to have series expansions valid in the limits of low or high frequency. In the former case we have

$$\Gamma = 1 - \tfrac{1}{2} i(\gamma - 1)\Delta \omega \tau - \tfrac{1}{2}(\gamma - 1)\Delta \, [1 - \tfrac{1}{4}\Delta(3 + \gamma)](\omega \tau)^2 + \cdots \qquad (4.3.37)$$

while in the latter

$$\Gamma = (\gamma / \gamma_\infty)^{1/2} \{1 - \tfrac{1}{2} i(\gamma_\infty - 1)\Delta_\infty (\omega \tau)^{-1}$$
$$- \Delta_\infty [(1 + \Delta_\infty)(1 + \gamma_\infty) - \tfrac{1}{4}\Delta_\infty (\gamma_\infty - 1)^2] \, (\omega \tau)^{-2} + \cdots \}. \qquad (4.3.38)$$

Here γ_∞ is the value of γ_ω in the regime $\omega \tau \gg 1$, and $\Delta_\infty = C_{\text{int}} / (C_p - C_{\text{int}})$. In the dispersive region approximations to the 'exact' expressions are also found to be useful. In particular, since $\mu / 2\pi$ is always small compared with unity, the inverse of equation (4.3.33) may be developed as a series so that

$$(u/u_0)^2 = \text{Re}(\gamma_\omega/\gamma)[1 + 3(\mu/2\pi)^2 + \cdots] \qquad (4.3.39)$$

$$\mu = \pi[\text{Im}(\gamma_\omega)/\text{Re}(\gamma_\omega)][1 - (\mu/2\pi)^2 + \cdots]. \qquad (4.3.40)$$

The terms of the second and higher order in $\mu/2\pi$ are usually negligible compared with typical experimental imprecision which, at least in the case of the phase speed, is greatly increased in the dispersive region. We see that the dispersion is determined by the behaviour of $\text{Re}(\gamma_\omega/\gamma)$ and that the absorption per wavelength is $\pi \tan \phi$ in leading order; these are the results anticipated above in connection with equation (4.3.14).

In order to have a complete description of absorption and dispersion of sound in pure polyatomic gases it will be necessary to include the visco-thermal effects. In many cases where the relaxation times of the internal modes of interest are long, there is no significant translational dispersion, relaxation absorption dominates over the classical absorption, and the two may be taken as additive. However, this is not usually the case for ro-tational relaxation and even for vibrational relaxation in some molecules (e.g. n-C_4H_{10}, $\zeta_{\text{vib}} \approx 1$). We must therefore consider the manner in which viscothermal and relaxation effects combine in the regime where neither are small. Although relaxation is still expected to manifest itself mainly through a frequency-dependent compressibility, $\varkappa_T/\gamma_\omega$, a complication arises from the possible frequency dependence of the thermal conductivity in a polyatomic gas.

According to the kinetic theory of dilute polyatomic gases, the ratio $\varkappa/\eta c_V$ may be written

$$\varkappa/\eta c_V = \mathfrak{F}_t + \sum_n \mathfrak{F}_n c_n/c_V \qquad (4.3.41)$$

where $\mathfrak{F}_t$ is a factor arising from the translational modes and the factors $\mathfrak{F}_n$ arise from the heat transported by each of the internal modes. There are several theoretical treatments of the factors $\mathfrak{F}_t$ and $\mathfrak{F}_n$ of varying degrees of complexity [12–14]. According to the simplified arguments put forward by Eucken [12], $\mathfrak{F}_t$ has the same value, 2.5, as for the monatomic gases, and the transport of heat by the internal modes is a simple diffusion process. From this it was shown that[3]

$$\varkappa/\eta c_V(\omega) = (9\gamma_\omega - 5)/4. \qquad (4.3.42)$$

Since the internal heat capacity is subject to relaxation, the thermal con-ductivity in this model is complex and frequency dependent—a prediction that is not altered by more sophisticated theories. For present purposes, it is not profitable to pursue more accurate prediction of the dynamic thermal conductivity.

One approach to the problem, then, is to evaluate the solutions of the hydrodynamic equations using a frequency-dependent heat capacity ratio and a frequency-dependent thermal conductivity—the latter obtained from an expression such as Eucken's. Unfortunately this approach tends to obscure the interesting relaxation effects, rather than effect any separation of them from the viscothermal processes.

Although there is no exact theory that allows complete separation of translational and relaxation effects in real fluids, we mention one model that does have this property. This is the so-called Becker gas, characterized by $\varkappa/\eta c_V(\omega) = 4\gamma_\omega/3$ and treated in the Navier–Stokes approximation [4]. It is found that the Navier–Stokes equation factorizes for a Becker gas and that its propagational solution may be written

$$\Gamma = \Gamma_t \Gamma_r = \Gamma_t (\gamma/\gamma_\omega)^{1/2} \qquad (4.3.43)$$

where Γ_t is the reduced propagation constant that would arise if internal relaxation were completely absent, and $\Gamma_r = (\gamma/\gamma_\omega)^{1/2}$ is the value that would arise with internal relaxation but in the absence of any translational effects. The static value of the ratio $\varkappa/\eta c_V$ for many polyatomic gases does in fact lie within a few per cent of the value appropriate to the Becker gas, and solutions of the Navier–Stokes equation for other models, such as the Eucken equation, are almost indistinguishable. These observations are probably useful in interpreting data at frequencies where the translational dispersion is small. Equation (4.3.43) has also been applied *ad hoc* with Γ_t estimated from the Burnett equation [15]. Unfortunately, both the Navier–Stokes and the Burnett theories are unreliable at very high frequencies. However, the kinetic models mentioned in §4.2 have been extended to relaxing diatomic gases with some considerable success [16]. The agreement between a seven-moment model and the experimental results for O_2 and N_2 was found to be excellent across the full frequency range studied $(0.001 \leqslant f/f_c \leqslant 1)$ when the rotational collision number was included as the single adjustable parameter [16]. This would appear to be the only approach capable of yielding accurate results at frequencies above a few per cent of f_c.

4.3.5   Vibrational relaxation

Usually the collision numbers characterizing vibrational relaxation are quite large and the experiments are conducted under conditions where translational and rotational dispersions are small. In that case the dispersion is dominated by vibrational relaxation and all of the effects of rotational relaxation can be included with the classical absorption through a bulk viscosity $\eta_b = [(\gamma - 1)\rho u^2] \tau_{rot} \Delta_{rot}$. We further assume the case of strong coupling of vibrational modes so that the total vibrational contribution to the heat capacity C_{vib} relaxes as one with a single apparent relaxation time τ_{vib}.

The heat capacity ratio will then be given by equation (4.3.14) with

$\tau = \tau_{vib}$ and, dropping the subscript, we have from the leading term of (4.3.39):

$$\left(\frac{u}{u_0}\right)^2 = 1 + \frac{\omega^2 \tau^2 (\gamma - 1)(1 - \gamma\Delta)\Delta}{[1 + \omega^2 \tau^2 (1 - \gamma\Delta)^2]}. \tag{4.3.44}$$

The dispersion relation in this form has been widely studied in both gases and liquids [17, 18]. In the former, the product τp is nearly independent of pressure and it is commonplace to perform the measurements at a constant frequency over a range of pressures: the variable ω is then replaced by ω/p and its coefficient τ by τp. This entails some complications: both u_0 and Δ are functions of the pressure in a real gas and account of this must be taken. Usually corrections are applied to reduce the results to that of the perfect gas. Once such corrections are made, it is a simple matter to obtain τp either by fitting the results to equation (4.3.44) or by plotting $(u/u_0)^2$ against $\log(\omega/p)$ and determining the point of inflexion. A better approximation is obtained if the results are analysed on the basis of $\tau\rho$ being constant rather than τp. Of course, variable frequency measurements at constant density are more easily interpreted.

Absorption measurements have also been used to determine vibrational relaxation times through the leading term of equation (4.3.40) in terms of which

$$\mu = \frac{\pi\omega\tau(\gamma - 1)\Delta}{[1 + \omega^2 \tau^2 (1 - \Delta)(1 - \gamma\Delta)]}. \tag{4.3.45}$$

Corrections are first applied for the classical (and possibly rotational) absorption before comparison with equation (4.3.45). At the angular frequency ω_{max} where the absorption per wavelength is maximum we have

$$\mu_{max} = \frac{\pi(\gamma - 1)(\Delta/2)}{[(1 - \Delta)(1 - \gamma\Delta)]^{1/2}} \tag{4.3.46}$$

which can be used to determine the value of Δ.

As an example, figure 4.4 shows the speed and absorption of sound in CF_4 at 295 K as functions of f/p [20]. The curves drawn in the figure are based not on a fit to these data but on the values $\gamma = 1.159$, $\Delta = 0.45$ and $\tau p = 73$ ms Pa obtained in much more recent work using frequencies well below the region of significant dispersion [21]. With the exception of a few of the absorption measurements, the agreement shown in the figure is rather good. In this case the whole vibrational contribution to the heat capacity evidently relaxes in unison.

Acoustic measurements on fluids at frequencies below the dispersive region are a highly accurate means of studying thermodynamic properties. When this is the objective of the measurements, the 'tail' of the dispersion curve is a nuisance to be corrected for. Fortunately, the effects of relaxation at low frequencies are much greater in the case of absorption than they are for the phase speed with $\mu = O(\omega\tau)$ and $[(u/u_0) - 1] = O(\omega^2\tau^2)$.

Thus, measurements of μ can be used to determine τ, through the imaginary part term in the series expansion of Γ (equation (4.3.37)), and that value then used to make the small correction from u to u_0 using the real part of the series.[4] In a few cases, notably simple diatomic molecules such as nitrogen, the vibrational relaxation time in the gas phase at low pressures is so long that it is easier to perform the measurements at frequencies far above the dispersive zone. Here too the dispersive effect of relaxation is small (of order $(\omega\tau)^{-2}$) while the effect on absorption is much larger (of order $(\omega\tau)^{-1}$); the results can be corrected to a hypothetical infinite-frequency speed of sound given by

$$u_\infty^2 = u_0^2(\gamma_\infty/\gamma). \qquad (4.3.47)$$

That quantity contains exactly the same information about residual (as opposed to perfect-gas) thermodynamic properties as does u_0^2.

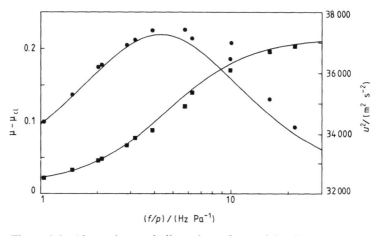

$(f/p) / (\text{Hz Pa}^{-1})$

Figure 4.4 Absorption and dispersion of sound in CF$_4$: $\bullet$, $\mu - \mu_{cl}$ experimental absorption per wavelength μ [20] minus calculated classical absorption $\mu_{cl} = \alpha_{cl}\lambda$; $\blacksquare$, experimental u^2 [20] after correction for gas imperfections using $\beta_a = -120 \text{ cm}^3 \text{mol}^{-1}$ [21]; —————, calculated from equations (4.3.44) and (4.3.45) using data from [21].

4.3.6 Rotational relaxation

As has already been indicated, rotational relaxation is characterized by small collision numbers. For this reason, the dispersive region occurs only at very high frequencies where translational relaxation is well under way. Here interpretation of the results rests on the extent to which the relaxation and translational effects can be combined. However, useful information can often be obtained at frequencies, well below that domain, where the dispersion is but slight and absorption measurements are most useful. Here the relaxation and classical absorptions are additive and the ratio α_r/α_t is given by

$$\frac{\alpha_r}{\alpha_t} = \frac{4\gamma(\gamma - 1)\Delta}{5[(\gamma - 1)(x/\eta c_p) + 4/3]} \zeta_{rot}. \qquad (4.3.48)$$

For example, in a perfect diatomic Becker gas ($x/\eta c_p = 4/3$, $\gamma = 7/5$, $\Delta = 2/7$) this ratio is $0.0686\zeta_{rot}$. Although quite accurate absorption measurements are required to measure small rotational collision numbers using frequencies such that $\omega\tau \ll 1$, the theory is well understood in that regime. Where necessary, account may be taken of absorption due to vibrational relaxation, and this presents no difficulties provided that $\tau_{vib} \gg \tau_{rot}$ and $\tau_{vib}^{-1} \ll \omega \ll \tau_{rot}^{-1}$.

We illustrate these points by reference to the experimental results for oxygen near to 300 K [15]. In this case the vibrational contribution to the heat capacity is small while ζ_{vib} is very large (approximately 10^6). Thus the only significant relaxation effects at frequencies above a few hertz arise from the rotational modes. In figure 4.5 the experimental attenuation (as $\alpha\lambda_0/2\pi$) is plotted against the reduced frequency $\omega\tau_c/2\pi$ in the region of small dispersion. The dotted line shows $-\text{Im}(\Gamma_t)$ calculated from the Navier–Stokes theory using the static values of γ_ω and x. The residual absorption corresponding to relaxation was fitted with the expression

$$\alpha_r\lambda_0/2\pi = \tfrac{1}{2}(\gamma - 1)\Delta\zeta_{rot}\omega\tau_c \qquad (4.3.49)$$

which follows from equation (4.3.37) in the absence of significant vibrational contributions. This procedure gave $\zeta_{rot} = 3.91 \pm 0.08$.[5] Figure 4.5 shows the excellent agreement with experiment that is achieved at low frequencies using this single adjustable parameter. In figure 4.6, the speed and absorption of sound in oxygen is plotted at frequencies up to $O(f_c)$ and compared with the seven-moment kinetic model of Hanson *et al* [16] which incorporates the value of ζ_{rot} obtained from the low-frequency sound absorption. Also shown are results calculated from the Navier–Stokes equation combined with the Becker model. The agreement between experiment and the predictions of the kinetic model is excellent across the entire frequency range and contrasts with the poor performance obtained at high frequencies from the Navier–Stokes equation.

It should be remarked that the assumption of a single relaxation time for the rotational modes is implicit in these statements. Molecules other than hydrogen are expected to behave as classical rigid rotors under accessible conditions because the spacing of rotational levels is small compared with the thermal energy of the molecules. In that case a single relaxation time is to be expected. However, the *ortho* and *para* states of H_2 (and D_2) exhibit significantly different relaxation times corresponding to the predominant rotational transitions involved [18]. In these exceptional

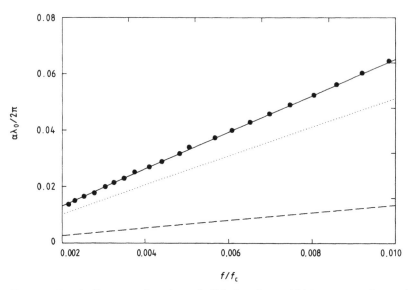

Figure 4.5 $\alpha\lambda_0/2\pi$ as a function of f/f_c for O_2 at 300 K: •, experimental results of Greenspan [15]; ······, classical absorption; — — — —, relaxation absorption calculated using $\zeta_{rot} = 3.91$; ——————, relaxation plus classical absorption.

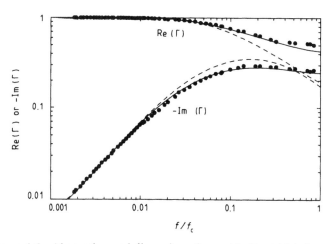

Figure 4.6 Absorption and dispersion of sound in O_2 at high frequencies: •, experimental results of Greenspan [15]; ——————, seven-moment kinetic model [16]; — — — —, Navier–Stokes theory plus Becker model with $\zeta_{rot} = 3.91$.

cases the spacing of rotational energy levels is comparable with the translational energies of the molecules. Consequently, quantum mechanical restrictions (which are different for *ortho* and *para* molecules) become an important factor governing the transfer of energy between rotational and translational modes.

4.4 Chemical and Structural Relaxation

In addition to the influence of thermal relaxation, chemical and structural rearrangements in the fluid may contribute to absorption and dispersion of sound. Clearly, if the extent of one or more chemical reactions can be displaced by the action of sound then the thermophysical properties will no longer be those of a phase of fixed composition. At frequencies low enough for local chemical equilibrium to be maintained at all times, the consequences of chemical reactions can be incorporated on purely thermodynamic grounds. At higher frequencies, the finite reaction rate will cause the extent of reaction to lag behind that which would be achieved if local equilibrium were maintained. This will result in absorption and dispersion similar to that found for systems undergoing thermal relaxation. Structural rearrangements can also be treated within the framework of chemical effects even though the precise nature of the structural changes may be unknown. In fact, the observable quantities (u and μ as functions of frequency) do not provide sufficient information to discriminate between thermal, chemical and structural relaxations. Only the magnitude of the effects and their dependence on temperature, pressure and overall composition provide any basis on which to assign absorption and dispersion to any one of these mechanisms.

Although it is possible to write down and solve the hydrodynamic equations for the general case of sound propagating through a fluid mixture in which one or more chemical reactions may be periodically displaced from equilibrium, we shall not attempt to include all possible effects simultaneously. Instead the various mechanisms contributing to the absorption and dispersion of sound will be assumed separable so that the transport of heat, momentum and mass may be ignored here. This will only be unsatisfactory if two or more mechanisms lead to large effects in the same frequency range.

4.4.1 Thermodynamic relations

We begin with a few simple relations which show the effect of a chemical reaction on the zero-frequency speed of sound. A general reaction may be written as

$$\sum_B \nu_B B = 0 \tag{4.4.1}$$

where B denotes a chemical substance with stoichiometric number ν_B. The extent of reaction ξ is defined for this general reaction by the relation $(\partial n_B / \partial \xi) = \nu_B$ in which n_B denotes the amount of substance B present in the mixture. It is also useful to define the specific affinity a of the reaction by the relation

$$a = -(\partial g / \partial \xi)_{p,T} \tag{4.4.2}$$

where g is the specific Gibbs function; the fundamental equation for a change of the state of a closed phase may then be written

$$dg = -s\,dT + \rho^{-1}\,dp - a\,d\xi \tag{4.4.3}$$

where s denotes specific entropy.

We will be interested in the values of the quantities c_p, $(\partial \rho / \partial p)_T$ and $(\partial \rho / \partial T)_p$ from which we can compute the sound speed through equation (1.2.2). If local equilibrium is maintained at all times then the affinity is always zero and the specific heat capacity at constant pressure is given by

$$c_p = (\partial h / \partial T)_{p,a} = (\partial h / \partial T)_{p,\xi} + (\partial h / \partial \xi)_{p,T}(\partial \xi / \partial T)_{p,a} \tag{4.4.4}$$

where h is specific enthalpy. Using the identity

$$(\partial \xi / \partial T)_{p,a} = \phi^{-1}(\partial a / \partial T)_{p,\xi} \tag{4.4.5}$$

where $\phi = -(\partial a / \partial \xi)_{p,T}$, together with the relation

$$\begin{aligned}(\partial a / \partial T)_{p,\xi} &= -[\partial (\partial g / \partial T)_{p,\xi} / \partial \xi]_{p,T} \\ &= (\partial s / \partial \xi)_{p,T}\end{aligned} \tag{4.4.6}$$

the derivative $(\partial \xi / \partial T)_{p,a}$ may be expressed as $\phi^{-1}(\partial s / \partial \xi)_{p,T}$. Notice that, since the Gibbs function is a minimum at a stable equilibrium, ϕ is positive. In addition, $(\partial s / \partial \xi)_{p,T} = (\partial h / \partial \xi)_{p,T} / T$ along the equilibrium path so that the specific heat capacity becomes

$$\begin{aligned}c_p &= (\partial h / \partial T)_{p,\xi} + [(\partial h / \partial \xi)_{p,T}^2 / T\phi] \\ &= (\partial h / \partial T)_{p,\xi}(1 + \Delta_1)\end{aligned} \tag{4.4.7}$$

where $(\partial h / \partial T)_{p,\xi}$ is the heat capacity that would apply if the reaction were frozen and Δ_1 gives the fractional enhancement arising from the chemical degree of freedom. Using analogous arguments, it can be shown that

$$\begin{aligned}(\partial \rho / \partial p)_{T,a} &= (\partial \rho / \partial p)_{T,\xi} - [(\partial \rho / \partial \xi)_{p,T}^2 / \rho^2 \phi] \\ &= (\partial \rho / \partial p)_{T,\xi}(1 + \Delta_2)\end{aligned} \tag{4.4.8}$$

and

$$\begin{aligned}(\partial \rho / \partial T)_{p,a} &= (\partial \rho / \partial T)_{p,\xi} + [(\partial h / \partial \xi)_{p,T}(\partial \rho / \partial \xi)_{p,T} / T\phi] \\ &= (\partial \rho / \partial T)_{p,\xi}(1 + \Delta_3).\end{aligned} \tag{4.4.9}$$

Since Δ_1 and Δ_2 are necessarily positive along an equilibrium path, we see that the effect of the chemical reaction is always to increase the heat capacity and isothermal compressibility above the values that would apply with the composition frozen. Only Δ_3 is affected by the signs of the enthalpy and density changes accompanying reaction.

Formally, the theory can be applied also to structural rearrangements in the fluid. To do this, one need only generalize the 'order parameter' ξ such that it measures the extent of the structural change. The thermodynamic treatment is then identical to that given above.

4.4.2 Kinetic relations

Since the process will have a finite rate constant, equilibrium cannot be maintained during the acoustic cycle and the order parameter ξ will differ from, but relax towards, the value ξ^{eq} that would apply if equilibrium were maintained. For small displacements, we assume that equilibrium is approached in accordance with the first-order equation

$$(\partial\xi/\partial t) = -(\xi - \xi^{eq})/\tau \qquad (4.4.10)$$

with relaxation time τ. When the reaction is displaced from equilibrium by the simple-harmonic fluctuation imposed by the sound field, we see, following arguments analogous to those employed in connection with thermal relaxation, that the dynamic value of the order parameter will be $\xi^{eq}/(1 + i\omega\tau)$. The effective values of c_p, $(\partial\rho/\partial p)_T$ and $(\partial\rho/\partial T)_p$ at angular frequency ω will therefore be given not by equations (4.4.7–9) but by relations in which the contributions corresponding to the change in ξ are modified by the factor $(1 + i\omega\tau)^{-1}$:

$$c_p = (\partial h/\partial T)_{p,\xi}[1 + \Delta_1(1 + i\omega\tau)^{-1}] \qquad (4.4.11)$$

$$(\partial\rho/\partial p)_T = (\partial\rho/\partial p)_{T,\xi}[1 + \Delta_2(1 + i\omega\tau)^{-1}] \qquad (4.4.12)$$

$$(\partial\rho/\partial T)_p = (\partial\rho/\partial T)_{p,\xi}[1 + \Delta_3(1 + i\omega\tau)^{-1}]. \qquad (4.4.13)$$

4.4.3 Speed and absorption of sound

Usually the relaxation parameters Δ_1, and Δ_2 and Δ_3 are small compared with unity and solutions correct just to leading order in these quantities are appropriate. The situation is similar to that for relaxation of internal degrees of freedom; the complex propagation constant k will again be equal to the dynamic value of $(\rho\chi_T/\gamma)^{1/2}$ but now both the isothermal compressibility and the ratio of heat capacities are complex and frequency dependent. It is useful to introduce a dimensionless propagation constant Γ_∞ defined in terms of the sound speed u_∞ that would apply if the chemical or structural process were frozen:

$$\Gamma_\infty = (ku_\infty/\omega) = (u_\infty/u)[1 - i(\mu/2\pi)]. \qquad (4.4.14)$$

Then since $\rho\chi_T/\gamma = [(\partial\rho/\partial p)_T - (T/\rho^2 c_p)(\partial\rho/\partial T)_p^2]$, we obtain the following expression for Γ_∞ from equations (4.4.11–13):

$$\Gamma_\infty = 1 + \tfrac{1}{2}[(\gamma_\infty - 1)\Delta_1 + \gamma_\infty\Delta_2 - 2(\gamma_\infty - 1)\Delta_3](1 + i\omega\tau)^{-1} \quad (4.4.15)$$

correct to leading order in the relaxation parameters. The sound speed and absorption are therefore given by

$$\frac{u}{u_\infty} = 1 - \left(\frac{[(\gamma_\infty - 1)\Delta_1 + \gamma_\infty\Delta_2 - 2(\gamma_\infty - 1)\Delta_3]}{2[1 + (\omega\tau)^2]}\right) \qquad (4.4.16)$$

and

$$\frac{\alpha\lambda_\infty}{2\pi} = i\omega\tau\left(\frac{[(\gamma_\infty - 1)\Delta_1 + \gamma_\infty\Delta_2 - 2(\gamma_\infty - 1)\Delta_3]}{2[1 + (\omega\tau)^2]}\right) \qquad (4.4.17)$$

where $\lambda_\infty = 2\pi(u_\infty/\omega)$. Since the second term on the right-hand side of equation (4.4.16) is necessarily positive (on thermodynamic grounds), the sound speed increases with frequency from u_0 at zero frequency to u_∞ as $\omega \to \infty$, passing through the midpoint at $\omega\tau = 1$. For a chemical reaction in dilute solution, the change in sound speed may be too small to detect but the absorption, which follows a curve indistinguishable from the case of thermal relaxation, is often measurable. In aqueous solution, where $(\gamma - 1)$ is small, chemical reactions are driven chiefly by the density fluctuations and the contribution of Δ_2 dominates the absorption and dispersion. This quantity can be obtained from the maximum value of the absorption per wavelength and used to infer the value of $(\partial\rho/\partial\xi)_{p,T}$. More generally, it is not possible to separate the effects of density and enthalpy changes accompanying chemical or structural changes and, for example, values of enthalpy changes for isomerization reactions obtained with the assumption $(\partial\rho/\partial\xi)_{p,T} = 0$ often deviate from those obtained by other means [22]. However, the frequency of maximum absorption does lead to a reliable measure of the rate constant in many cases.

References

[1] Chapman S and Cowling T G 1952 *The Mathematical Theory of Nonuniform Gases* (Cambridge: Cambridge University Press)
[2] Burnett D 1935 *Proc. London Math. Soc.* **39** 385; **40** 382
[3] Grad H 1949 *Commun. Pure Appl. Math.* **2** 331
[4] Greenspan M 1965 *Physical Acoustics* vol 2A ed. W P Mason (New York: Academic Press) pp 1–45
[5] Pekeris C L, Alterman Z, Finkelstein L and Frankowski K 1962 *Phys. Fluids* **5** 1608

[6] Sirovich L and Thurber J K 1965 *J. Acoust. Soc. Am.* **37** 329

[7] Greenspan M 1956 *J. Acoust. Soc. Am.* **28** 644

[8] Meyer E and Sessler G 1957 *Z. Phys.* **149** 15

[9] Aziz R A, Bowman D A and Luis C C 1967 *Can. J. Phys.* **45** 2079

[10] Baharudin B Y, Jackson D A, Schoen P E and Rouch J 1973 *Phys. Lett.* **46A** 39

[11] Baharudin B Y, Jackson D A and Schoen P E 1974 *Molecular Motion in Liquids, Proc. 24th Ann. Meet. of the Société de Chimie Physique* ed. J Lascombe (Dordrecht: Reidel) pp 597–604

[12] Eucken A 1913 *Phys. Z.* **14** 324

[13] Mason E A and Monchick L 1962 *J. Chem. Phys.* **36** 1622

[14] Viehland L A, Mason E A and Sandler S I 1978 *J. Chem. Phys.* **68** 5277

[15] Greenspan M 1959 *J. Acoust. Soc. Am.* **31** 155

[16] Hanson F B, Morse T F and Sirovich L 1969 *Phys. Fluids* **12** 84

[17] Herzfeld K F and Litovitz T A 1959 *Absorption and Dispersion of Ultrasonic Waves* (London: Academic Press)

[18] Lambert J D 1977 *Vibrational and Rotational Relaxation in Gases* (Oxford Clarendon)

[19] Cottrell T L and McCoubrey J C 1961 *Molecular Energy Transfer in Gases* (London: Butterworths)

[20] Beyers W H 1943 *J. Chem. Phys.* **11** 348

[21] Ewing M B and Trusler J P M 1989 *J. Chem. Phys.* **90** 1106

[22] Wyn-Jones E and Orville-Thomas W J 1966 *Molecular Relaxation Processes* Chemical Society Special Publication No. 20 (London: Academic Press) pp 209–44

[23] Kohler M 1949 *Z. Phys.* **127** 41

[24] Holmes R and Tempest W 1960 *Proc. R. Soc.* A **75** 898

[25] Huch R J and Johnson E A 1980 *Phys. Rev. Lett.* **42** 142

Notes

1. A Maxwellian gas is one composed of molecules that repel one another in proportion to the inverse fifth power of their separation.

2. We use the notation $E(T)$ to denote the energy stored by a mode when its characteristic temperature is T.

3. Eucken's equation is a fair approximation for most simple polyatomic gases under static conditions. Whatever validity the model possesses is retained for a relaxing fluid.

4. This requires knowledge of the relaxing heat capacity.

5. The value $\zeta_{rot} = 4.09$ given in the original work was obtained assuming Eucken's value of $\varkappa/\eta c_p$ (1.36); the result given here was recalculated using the experimental ratio $\varkappa/\eta c_p = 1.41$.

5

Generation and Detection of Sound

5.1 Introduction

Transducers for the generation and detection of sound are one of the most important components of experimental apparatus; they are also one of the most problematic. The requirements that transducers must satisfy before they are useful in the present context are quite different to those commonly encountered in audio engineering. For the kind of measurements described in this book, very small power outputs are required (usually less than 1 μW) and the main applications can be broadly categorized into three classes. In the first, we have wide-bandwidth devices that present a high acoustic impedance to the medium. Typically, such transducers form part of the walls of a resonator and we require the acoustic impedance to be large so that the transducers do not perturb the properties of the cavity too much. In the second class, we have narrow-bandwidth devices for use in fixed-frequency variable pathlength interferometers. These are typically operated at a mechanical resonance frequency where the intrinsic mechanical impedance is close to zero and the electrical input impedance is determined mainly by the impedance presented to the transducer by the acoustic load. Finally we have transducers for pulsed operation. These are also fixed-frequency devices but larger bandwidths are often required to achieve useful time-domain resolution.

Before continuing, it would be as well to include a few comments about the analogies between mechanical and electrical quantities. The concept of an acoustic impedance $Z_a = p_a/v$ relating acoustic pressure and fluid speed has already been introduced. It is most useful for wave motion in one dimension. When we come to deal with systems where the driving force is applied essentially at a point (or perhaps over a small area) and the motion is still primarily one dimensional, the mechanical impedance Z_m will be a more appropriate quantity. This is defined as the ratio F/v of the force F

and the speed v at the driving point. For a point mass driven by a simple-harmonic force $F \exp(i\omega t)$, but constrained by a stiffness force proportional to displacement and a resistive force proportional to speed, the mechanical impedance is given by

$$Z_m = F/v = R_m + i\omega m - is/\omega. \qquad (5.1.1)$$

Here, $v \exp(i\omega t)$ is the steady-state speed, R_m is the mechanical resistance, m is the mass and s is the stiffness (the inverse of the mechanical compliance). Clearly, this impedance is analogous with the electrical impedance when we interpret force like potential difference, speed like current, mechanical resistance like electrical resistance, mass like inductance and compliance like capacitance. Often we can then think of a mechanical system as a circuit comprised of lumped elements (point masses, mechanical resistors and compliances) that combine according to the mechanical equivalent of Kirchhoff's laws; however, we shall not push this analogy too far.

A transducer is a device for the conversion of electrical work into mechanical work or vice versa. It has a mechanical port where the motion takes place and an electrical port through which current may pass. When we consider the electrical or mechanical impedance of the device we must allow for coupling between these ports. For example, although the mechanical impedance presented by the diaphragm of a microphone is affected by such things as mass and stiffness, it is also affected by the current induced to flow through the electrical port by virtue of the diaphragm's motion. If the output of the microphone is connected to a resistive load then power will be dissipated there when a current flows; this will introduce damping at the mechanical port. The detailed nature of the coupling between electrical and mechanical ports will depend upon the operating principle of the transducer. Such interaction is exploited in single-transducer interferometers, where the mechanical impedance presented to the face of the transducer by a fluid column is inferred from measurements of the electrical input impedance of the transducer. Consequently, we shall be concerned with calculation of both the electrical and the mechanical impedance of loaded transducers. Often there will also be some trade off between signal strength and other requirements such as minimizing perturbations to the performance of a resonator; consequently, it will be useful also to have some guidance about the efficiency of various designs. Finally, we may well require the transducers to operate over a wide range of temperature and pressure while maintaining an acceptable signal-to-noise ratio; this requirement places severe constraints upon the range of materials from which practical devices may be fabricated.

This chapter covers both theoretical and practical aspects of transduction. In the next two sections, a few basic concepts about sources, impedance, radiation and vibrating diaphragms are discussed, while in the later sections design and operation of various types of transducer are discussed in detail.

5.2   Radiation from Point and Piston Sources

5.2.1   The simple source

In this section, the concept of the point source is developed and its radiation pattern calculated. The point source is a mathematical device, rather like the point mass, that may be viewed as a small pulsating solid sphere, the diameter of which is very much less than the wavelength of the sound that it generates in the surrounding medium. The utility of this mathematical device lies in the fact that any finite source may be thought of as an assemblage of point sources so that, once the radiation from a point source has been found, other more practical problems can be solved in a fairly simple manner.

Let us first consider a finite spherical source of mean radius a placed at the origin of the coordinate system. If the sphere is expanding and contracting in an isotropic and harmonic manner with angular frequency ω then the sound field generated in an unbounded medium will consist of an outward-going spherical wave with acoustic pressure

$$p_a(r, t) = (A/r)\exp[i(\omega t - kr)] \qquad (5.2.1)$$

where r is the distance from the centre of the sphere and k is the propagation constant.[1] Of course, this expression is meaningful only for values of $r > a$. Since the fluid velocity $v = (i/\omega\rho)\nabla p_a$, the constant A and the radial fluid speed v can be evaluated from the boundary condition at the surface of the sphere: $v(a, t) = v_0 \exp(i\omega t)$. The results are

$$p_a(r) = \rho u(S_\omega/4\pi a^2)(a/r)(1 - i/ka)^{-1} \exp[ik(a - r)] \qquad (5.2.2)$$

and

$$v(r) = (S_\omega/4\pi a^2)[(1 - i/kr)/(1 - i/ka)] (a/r)\exp[ik(a - r)] \quad (5.2.3)$$

the time dependence $\exp(i\omega t)$ now being assumed implicitly. The quantity S_ω used here is called the source strength. It is defined for a source of arbitrary shape whose surface vibrates with a single angular frequency ω, but not necessarily everywhere with the same amplitude and phase, by the integral, over the surface S of the object, of the outward-pointing normal surface speed. Thus,

$$S_\omega \exp(i\omega t) = \int_S v \cdot n \, dS \qquad (5.2.4)$$

where n is an outward-pointing unit vector normal to the surface. For the isotropically vibrating sphere this quantity is real and equal to $4\pi a^2 v_0$. Now, if we take the limits of equations (5.2.2) and (5.2.3) as the radius a tends to zero, keeping S_ω constant, then we obtain expressions for the acoustic pressure and radial fluid speed generated by a point source:

$$p_a(r) = i(k\rho u/4\pi r)S_\omega \exp(-ikr) \qquad (5.2.5)$$

and

$$v(r) = (1/4\pi r^2)(1 + ikr)S_\omega \exp(-ikr). \qquad (5.2.6)$$

It will be worth while examining these equations in a little detail before proceeding. The acoustic pressure takes the form of a simple outward-going spherical wave whose amplitude drops off as $1/r$. The fluid speed, however, has two components. The first, $(ik/4\pi r)S_\omega \exp(-ikr)$, is also a simple spherical wave propagating outwards from the source in phase with the acoustic pressure but out of phase with the surface speed of the source; it also decays like $1/r$. At distances large compared with the wavelength λ, this term dominates the sound field and the relation between p_a and v approaches that for a plane wave: $p_a/v = \rho u$. This region is called the far field. Here, the mean-squared energy density, $w = \frac{1}{2}\rho\,|\,v^2\,| + \frac{1}{2}\varkappa s\,|\,p_a\,|^2$, is just $|\,p_a\,|^2/\rho u^2$; the energy is radiant, moving out from the source with speed u and intensity uw. The second component of v, $(1/4\pi r^2)S_\omega \exp(-ikr)$, is out of phase with the acoustic pressure and it dominates in the near field $r \ll \lambda$. Here, there is an additional mean-square energy density of $\frac{1}{2}\rho(S_\omega/4\pi r^2)^2$ and, since the energy of the near field contained *outside* a sphere of radius r_m drops off as $1/r_m$, we see that power is not radiated outwards from the source by this term. Instead, this component of the velocity describes fluid that is simply moved back and forth in phase with the motion of the source.

The point source is one member of a class of sources of arbitrary shape known as simple sources. A simple source is one with dimensions small in comparison with the wavelength, the surface of which executes simple-harmonic monofrequency vibrations. The amplitude and phase of these vibrations may be an arbitrary function of position over the surface of the object but one can show that the far field generated by the source is identical with that of any other simple source, including a point source, of the same strength. This will not be true of the near field.

5.2.2   Impedance of a point source

The reaction of the fluid back on to the source is described by the mechanical radiation impedance $Z_r = F/v_0$, where $F = 4\pi a^2 p_a(a)$ is the force acting on the fluid and v_0 is the surface speed. For a finite spherical source we have from equations (5.2.2) and (5.2.3):

$$Z_r = i\rho u (4\pi a^2) ka/(1 + ika) \qquad (5.2.7)$$

and for a point source this reduces to

$$Z_r = i\rho u (4\pi a^2) ka (1 - ika). \qquad (5.2.8)$$

The power radiated by the source is equal to the real part of Z_r multiplied by the square of the surface speed; for a point source this is given by

$$P = \rho u S_\omega^2 k^2 / 4\pi \qquad (5.2.9)$$

and is proportional to the square of the frequency. Any practical source, such as a coil-driven diaphragm, will also have an intrinsic mechanical impedance of its own. The force applied to the driving system and the resulting velocity will therefore be in the ratio determined by the sum of the intrinsic impedance and the radiation impedance.

Often we will prefer the dimensionless specific acoustic impedance $z = p_a/\rho u v$ to the mechanical impedance. For the simple point source the specific acoustic radiation impedance is

$$z_r = (ka)^2 + ika. \qquad (5.2.10)$$

5.2.3 The Green function

The sound field produced by a point source may also be expressed in terms of a Green function. Generalizing to the case of a point source located at point r_0, not necessarily coincident with the origin of the coordinate system, the field at an observation point r may be written

$$\Psi_\omega(r \mid r_0, t) = S_\omega g_\omega(r \mid r_0) \exp(i\omega t) \qquad (5.2.11)$$

where $\Psi_\omega(r \mid r_0, t)$ is the velocity potential,

$$g_\omega(r \mid r_0) = \frac{1}{4\pi R} \exp(-ikR) \qquad (5.2.12)$$

is the Green function for free space and $R = |r - r_0|$ is the distance from the source. As usual, the acoustic pressure is equal to $\rho (\partial \Psi / \partial t)$ and the fluid velocity is $-\nabla \Psi$. The Green function has several important features. For a start it is symmetric with respect to exchange of the source point and the measurement point, and therefore conforms to the principle of reciprocity. In addition, it is a solution of the inhomogeneous equation

$$(\nabla^2 + k^2) g_\omega(r \mid r_0) = -\delta(r - r_0) \qquad (5.2.13)$$

and therefore a solution of the homogeneous wave equation

$\nabla^2 g_\omega + k^2 g_\omega = 0$ everywhere except at the source point. To show this, one integrates both sides of equation (5.2.13) in the coordinates of r over the volume of a small sphere of radius b centred on r_0. In view of the properties of the delta function, the right-hand side evaluates to -1. On the left-hand side, the volume integral of $k^2 g_\omega$ approaches $(kb)^2$ for small b and therefore vanishes as $b \rightarrow 0$. The remaining term may be evaluated using Gauss' law to reduce the volume integral of $\nabla^2 g_\omega$ to the surface integral of ∇g_ω:

$$\int_0^b \nabla^2 g_\omega 4\pi R^2 \, dR = 4\pi b^2 (\partial g_\omega / \partial R)_{R=b} = -(1 + ikb)\exp(-ikb) \quad (5.2.14)$$

which approaches -1 when $b \rightarrow 0$ as required.

The Green function of equation (5.2.12) applies only to the unbounded medium. In order to allow for the presence of boundaries located within a finite distance of the source it will be necessary to add to $g_\omega(r \,|\, r_0)$ some function that is both a solution of the homogeneous wave equation and chosen to satisfy the boundary conditions. This additional function, which describes the reflected waves, may be simple or complicated depending upon the geometry but, whatever it is, the new Green function $G_\omega(r \,|\, r_0)$ is, like $g_\omega(r \,|\, r_0)$, a solution of the inhomogeneous equation (5.2.13). These observations form the basis of the methods used in Chapter 3 to construct a description of a cavity driven by a point source of sound.

5.2.4 The baffled point source

An example of the effect of a boundary is afforded by the case of a simple point source located a distance d from a plane solid surface. If the surface is perfectly rigid then the boundary condition there is that the normal component of the fluid velocity must vanish everywhere on the surface. The composite wave field that arises in this case can be thought of as the combination of waves travelling directly to the observation point from the actual source and waves travelling from an imaginary image source placed a distance d behind the surface such that a line joining the real source and its image lies normal to the surface. This combination satisfies the boundary conditions when the strengths of the source and its image are identical. In the limit $d \rightarrow 0$, the wave reflected from the surface is everywhere in phase with that coming directly from the source. Consequently, the acoustic pressure generated by a point source located on a solid plane boundary is exactly twice that of the same source in free space; the intensity is increased fourfold and the radiation resistance and power output are both doubled. In this case, the source is said to be baffled.

5.2.5   *The baffled piston source*

As indicated earlier the point source, which is of little practical value in itself, is a useful mathematical device because it allows the field from a distributed source to be derived. We shall have cause several times in the following sections to refer to the properties of transducers whose active elements approximate to a vibrating plane circular piston. When such a source is mounted on a rigid baffle of infinite extent the radiation pattern and impedance may be obtained with relative ease. This situation will be useful as a model.

Let the baffle lie in the x, y plane at $z = 0$ with the piston, of radius b, centred on the origin, and let the measurement point r be located by the polar coordinates r and θ, where r is the distance from the origin and θ is the angle made with the z axis. The piston is assumed to vibrate in the z direction with speed $v_0 \exp(i\omega t)$ and we divide its surface into infinitesimal elements, of area dS, each of which acts like a baffled point source of strength $v_0\, dS$. The position r_0 of each of these elementary sources may be given in terms of polar coordinates r_0 and θ_0, where θ_0 is the angle measured from the x axis. This system of source and observation point coordinates is illustrated in figure 5.1. The acoustic pressure $p_a(r \mid r_0)$ generated by each elementary source will be $(ik\rho u/2\pi R)\exp(-ikR)v_0\, dS$ where, as before, $R = |r - r_0|$ and the time dependence $\exp(i\omega t)$ is assumed implicitly. The total acoustic pressure is therefore given by the integral

$$p_a(r, \theta) = i(\rho u k v_0/2\pi) \int_0^{2\pi} \int_0^b \frac{\exp(-ikR)}{R}\, r_0\, dr_0\, d\theta_0. \qquad (5.2.15)$$

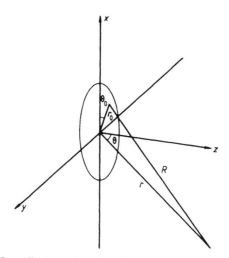

Figure 5.1   Coordinate system for piston source set in a plane baffle.

When $\theta = 0$, $R = (r_0^2 + r^2)^{1/2}$ independent of θ_0 and the integrand is a perfect differential: $(ikr_0/R)\exp(-ikR) = (d/dr_0)\exp(-ikR)$. Consequently, the acoustic pressure along the z axis may be evaluated very simply with the result

$$p_a(r, 0) = \rho u v_0 \exp(-ikr)[[1 - \exp\{ikr[1 - (1 + (b/r)^2)^{1/2}]\}]]$$

$$\xrightarrow[r/b \to \infty]{} i\tfrac{1}{2}\rho u v_0 kb(b/r)\exp(-ikr)$$

$$\xrightarrow[r/b \to 0]{} \rho u v_0 [\exp(-ikr) - \exp(-ikb)\exp(-ikr^2/2b)]. \qquad (5.2.16)$$

This is illustrated in figure 5.2 for two values of kb. Like a spherical wave, the acoustic pressure falls off as $1/r$ at large distances (great in comparison with the radius of the piston). However, the near field is more complicated and exhibits strong interference effects such that $|p_a|$ passes through extrema when r/b satisfies

$$ikr\{[1 + (b/r)^2]^{1/2} - 1\} = m\pi. \qquad (5.2.17)$$

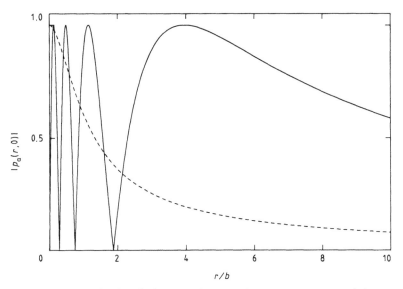

Figure 5.2 Magnitude of the on-axis acoustic pressure generated by a plane circular piston set in an infinite baffle: ————, $kb = 8\pi$ ($\lambda = b/4$); – – – –, $kb = \pi$ ($\lambda = 2b$).

Even values of the integer m correspond to minima (which are zeros) while odd values correspond to maxima, and the roots of equation (5.2.17) are

$$r_m/b = (b/m\lambda) - (m\lambda/4b). \qquad (5.2.18)$$

Approaching the source along the axis from far away, the acoustic pressure at first increases but, when $\lambda < 2b$, it passes through a maximum at r_1 and drops to zero at r_2. There are a number of additional extrema up to a maximum value of m not greater than $2b/\lambda$. The location r_1 of the most distant maximum, or λ when it is greater, can be taken as a rough measure of the extent of the near field. Note that for wavelengths greater than the diameter of the piston interference effects no longer play an important role and the extrema disappear. On the other hand, when the diameter of the piston is large compared with the wavelength, the interference is strong and essentially plane waves are produced close to the source. This can be shown by a rigorous treatment of equation (5.2.15); it can also be inferred from the behaviour of the on-axis acoustic pressure in the regime $r \ll b \gg \lambda$. In that case, the final line of equation (5.2.16) applies and the term containing $\exp(-ikb)$, which has the factor $\exp(-\alpha b)$, is rendered negligible by the absorption over distance of order b so that $p_a(r, 0) = \rho u v_0 \exp(-ikr)$. Since this attenuates only on account of the factor $\exp(-\alpha r)$, unaffected by geometric terms, the near field must consist of essentially plane waves propagating outwards in a nearly parallel beam of radius somewhat less than b.

Exact evaluation of the field at any angle and distance is rather complicated but one can easily obtain an expression that is valid in the far field when r is much greater than the radius b of the piston. In that case, one may neglect terms of order b/r or smaller and obtain the following approximations for R and $R^{-1} \exp(-ikR)$:

$$R \approx r - r_0 \sin \theta \cos \theta_0. \tag{5.2.19}$$

$$R^{-1} \exp(-ikR) \approx r^{-1} \exp(-ikr)\exp(ikr_0 \sin \theta \cos \theta_0). \tag{5.2.20}$$

The acoustic pressure in the far field is therefore given by

$$p_a(r, \theta) \approx i\rho u v_0 (k/2\pi r)\exp(-ikr) \int_0^{2\pi} \int_0^b \exp(ikr_0 \sin \theta \cos \theta_0)r_0 \, dr_0 \, d\theta_0 \tag{5.2.21}$$

and, making use of the standard integrals (see Appendix 1)

$$\int_0^{2\pi} \exp(iz \cos w) \, dw = 2\pi J_0(z) \tag{5.2.22}$$

and

$$\int J_0(az)z \, dz = (z/a)J_1(z) \tag{5.2.23}$$

the result is

$$p_a(r, \theta) = i\tfrac{1}{2} \rho u v_0 kb \left(\frac{b}{r}\right)\exp(-ikr)\left(\frac{2J_1(kb \sin \theta)}{kb \sin \theta}\right). \tag{5.2.24}$$

This shows that the acoustic pressure $p_a(r, \theta)$ in the far field is equal to the product of the on-axis value, $p_a(r, 0)$, and the angular-dependent term $H(v) = [2J_1(kb \sin \theta)/kb \sin \theta]$ illustrated as a function of $v = kb \sin \theta$ in figure 5.3. This reveals that there are nodes in the acoustic pressure on the conical surfaces with angles $\theta_n > 0$ that satisfy

$$kb \sin \theta_n = \psi_{1n} \qquad (5.2.25)$$

where ψ_{1n} are the zeros of the first-order cylindrical Bessel function. The first few values of ψ_{1n} are given in table 5.1 (note, however, that ψ_{11} does not lead to a valid solution; both numerator and denominator of $H(\nu)$ vanish as $kb \sin \theta \to 0$ but their ratio approaches unity).

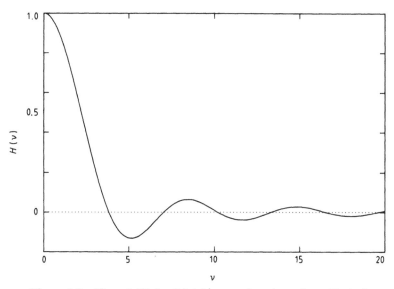

Figure 5.3 Plot of $H(\nu) = 2J_1(\nu)/\nu$ as a function of $\nu = kb \sin \theta$.

Table 5.1 Zeros of $J_1(z)$.

n	1	2	3	4	5
ψ_{1n}	0	3.83	7.02	10.17	13.32

$$\psi_{1n} = \chi_{0n}; \ \psi_{1n} \xrightarrow[n \to \infty]{} (n - 3/4)\pi$$

At low frequencies, there are no roots of equation (5.2.25) and the radiation pattern is the isotropic one discussed above, but when kb exceeds ψ_{12} the first node begins to develop. Towards higher frequencies, the central lobe of the radiation pattern, contained within the first conical nodal surface, contracts into a narrow beam. At the same time other outer lobes develop but, as the angular function of figure 5.3 shows, their amplitude falls off rapidly as the order n increases. In figure 5.4 polar diagrams show the radiation pattern from a baffled piston source for several values of kb.

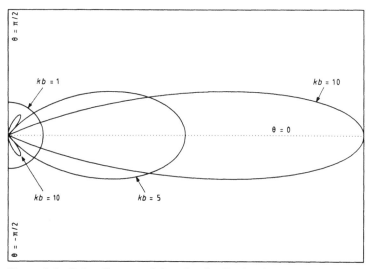

Figure 5.4 Polar diagram giving the distribution in angle of the RMS acoustic pressure at a fixed distance from a baffled piston source for various values of $kb = 2\pi b/\lambda$. A diffraction ring is evident when $kb = 10$.

In order to calculate the mechanical radiation impedance of the baffled piston source, we must evaluate the total reaction forces F on the face of the piston:

$$F = \int_0^b p_a(r, \pi/2) \ 2\pi r \ dr = i\rho u k v_0 \int_0^b \int_0^{2\pi} \int_0^b \frac{\exp(-ikR)}{R} \ r_0 \ dr_0 \ d\theta_0 r \ dr$$

$$(5.2.26)$$

where $R^2 = r^2 + r_0^2 - 2rr_0 \cos \theta_0$. This is a non-trivial integration but the mathematical details are unimportant here and the result may be written [1]:

$$Z_m = F/v_0 = \pi b^2 \rho u z_r \qquad (5.2.27)$$

where z_r is the specific acoustic radiation impedance given by

$$z_r = \left[1 - \left(\frac{1}{kb}\right)J_1(2kb)\right] + i\left[\left(\frac{4}{\pi}\right)\int_0^{1/2\pi} \sin(2kb \cos \alpha)\sin^2 \alpha \ d\alpha\right] \quad (5.2.28)$$

and illustrated in figure 5.4 as functions of kb. At very low or very high frequencies the following asymptotic forms apply:

$$z_r \rightarrow \tfrac{1}{2}(kb)^2 + i8kb/3\pi \qquad kb \rightarrow 0$$
$$z_r \rightarrow 1 + i2/\pi kb \qquad\qquad kb \rightarrow \infty. \qquad (5.2.29)$$

At low frequencies, the impedance is mainly reactive, owing to the mass of fluid being moved back and forth, while the power radiated varies like ω^2. This demonstrates the fact that sources with dimensions small compared with the wavelength are very inefficient radiators of acoustic energy. As we have seen, the radiation pattern has no preferred direction under these conditions. However, at high frequencies the acoustic radiation impedance is mainly resistive and close to the characteristic acoustic impedance of the medium, ρu. The piston is then coupled efficiently to the medium but the energy it radiates is concentrated mainly in a narrow axial beam.

If the baffle is removed so that we have a piston source in free space or, perhaps more realistically, attached to the end of a cylinder then the mathematical treatment of the problem becomes much more complicated. The only result that we shall make use of is the specific acoustic radiation impedance at low frequencies which is given approximately by [2]

$$z_r \approx \tfrac{1}{4}(kb)^2 + i0.6(kb).$$ (5.2.30)

In the absence of the baffle, the acoustic pressure is reduced by half and the intensity by three-quarters. However, the radiation pattern in the far field is now nearly spherical so the energy radiated at low frequencies is just half that from a baffled piston with the same speed amplitude and area. This accounts for the low-frequency limit of the radiation resistance which is half that for the baffled piston. At very high frequencies, the presence or absence of the baffle is of little importance because the interference between the radiation from different points on the piston concentrates the energy into a narrow beam.

Although useful, the direct summation of the fields from elementary sources is not always the best approach to solving problems. For example, another case of interest is that of a transducer that forms one complete end of a cylindrical interferometer. If we assume that it excludes plane piston-like motion so that only plane waves propagate in the cylinder then the radiation impedance may be obtained directly from the theory of the interferometer, as was done in Chapter 3. If we were interested in more complicated forms of transducer motion then a normal-mode analysis of the wave field inside the interferometer would be appropriate. In either case recourse to elementary sources is unnecessary.

5.2.6 Cavity impedance

One further task remains in this section. We will be interested later in acoustic resonators driven by small transducers whose dimensions are small compared with the wavelength of the sound they produce. It will be useful to evaluate the reaction of the medium on such a device so that we can estimate the behaviour of the source strength as the frequency is scanned through a resonance of the cavity. To do this, we first evaluate the impedance of a point source located inside the cavity. The acoustic pressure will be given by

$$p_\omega(r \mid r_0) = i\omega\rho S_\omega G_\omega(r \mid r_0) \qquad (5.2.31)$$

where $G_\omega(r \mid r_0)$ is the Green function for the cavity which may be expanded in terms of the normal modes as we did in Chapter 3:

$$G_\omega(r \mid r_0) = \sum_N \frac{\phi_N(r)\phi_N(r_0)}{V\Lambda_N(K_N^2 - k^2)} \qquad (5.2.32)$$

(for simplicity the eigenfunctions are taken here as purely real and the frequency dependence of the eigenvalues has been suppressed). If the source is a sphere of radius $a \ll \lambda$ then the reaction force on it will be the integral of $p_\omega(r \mid r_0)$, in the coordinates of r, over the surface of a sphere of radius a centred on r_0. But, for small values of a, the series for $G_\omega(r \mid r_0)$ converges only very slowly because there are an infinite number of normal modes contributing and more and more of them add in phase as $r \to r_0$. Consequently, most of the contribution close to the source comes from the terms of high order. The most important case in practice will be the one in which just one mode of the resonator (or one group of nearly degenerate modes) is excited with any substantial amplitude. This mode or group of modes, which dominate the sound field everywhere in the resonator except very close to the source, is then considered separately from the other terms in the normal-mode expansion of the Green function. Very close to the source, G_ω must approach in value the Green function for a point source in free space so that we may approximate the fluid velocity and acoustic pressure arising from the non-resonant terms in the normal-mode expansion of G_ω by those appropriate for the source in free space. Since we are assuming that $a \ll \lambda$, these will be in the ratio given by equation (5.2.10). At the surface of the source, the resonant mode makes a contribution to the fluid flow that is negligible by comparison with the very large near-field velocity but it does make a significant contribution to the acoustic pressure. This is given by

$$p_N = i\rho u k 4\pi a^2 v_0 \left[\phi_N^2(r_0) / V\Lambda_N(K_N^2 - k^2)\right] \qquad (5.2.33)$$

so that the specific acoustic radiation impedance approaches

$$z_r \approx ika + 4\pi a^2 ik \left[\phi_N^2(r_0) / V\Lambda_N(K_N^2 - k^2)\right] \qquad (5.2.34)$$

as $a \to 0$. Exactly at the Nth resonance, z_r reduces to

$$z_r \approx ika + (4\pi a^2 / kV) \left[\phi_N^2(r_0) / \Lambda_N\right] Q_N \qquad (5.2.35)$$

where Q_N is the quality factor of the resonance. Since the second term here is of order a^2, while the first is of order a, z_r does indeed approach the value ika for a source in free space when $a \to 0$, as it must. However, for *a* small but non-zero, there is an additional contribution from the resonant mode which is purely resistive at resonance and can be used to calculate the power dissipated in the cavity. The factor Q_N usually ensures that this term is much larger than the other 'second-order' contributions to the radiation impedance (i.e. other terms of order a^2) arising from the non-resonant modes. The mechanical radiation impedance is simply $S\rho u z_r$ so that the power radiated by the source is given by

$$P = S^2 |p_s|^2 / Z_m = S |p_s|^2 / \rho u z_r \tag{5.2.36}$$

where S is the area of the source and p_s is the acoustic pressure there.

A similar argument applies in the case of a small piston-like source mounted on the wall of the resonator. Thus, when the radius of the source b is much less than the wavelength of the sound, the radiation impedance approaches the sum of two terms, one for the baffled piston radiating into an infinite half space, and a second term for the resonant mode or modes. For a single mode excited, this becomes

$$z_r \approx i(8/3\pi)kb + \pi b^2 ik [\phi_N^2(r_0)/V\Lambda_N(K_N^2 - k^2)]$$
$$\xrightarrow[k \to K_N]{} i(8/3\pi)kb + (\pi b^2/kV)[\phi_N^2(r_0)/\Lambda_N]Q_N. \tag{5.2.37}$$

Note that this is not the impedance 'seen' by a sound wave reflecting from the surface of the transducer; it will see the sum of the radiation impedance and the intrinsic impedance of the piston itself, which will usually be very much greater.

As a numerical example, consider a source of diameter 3 mm driving an acoustic resonator of volume 1 l at a frequency of 5 kHz. If the speed of sound is 300 m s^{-1} ($k \approx 100 \text{ m}^{-1}$), $Q_N = 10^3$ and $\phi_N^2(r_0)/\Lambda_N = 1$ then $z_r \approx 0.07 + 0.13i$. The power radiated into a gaseous medium is extremely small: suppose that the density is 1.5 kg m^{-3} and that the sound pressure level at the face of the source transducer is 0.1 Pa; the power emitted is then about 0.5 nW.

5.3   Vibrating Elements

The behaviour of an electroacoustic transducer is determined largely by the properties of the vibrating element. Wide-band transducers usually employ membranes or thin metal plates as vibrating diaphragms because, when suitably mounted and driven, the displacement amplitude of these systems is independent of frequency in the region below the fundamental resonance frequency. Accordingly, in this section, the acoustic properties of membranes and plates are described in sufficient detail to allow a proper understanding of practical transducers.

5.3.1 The circular membrane

In the absence of damping, the transverse displacement $\xi(r)$ of a uniform circular membrane fixed at its rim and driven on one side by a uniform acoustic pressure $p_a \exp(i\omega t)$ is given by [3]

$$\xi(r) = \left(\frac{p_a}{k_M^2 \Im}\right)\left(\frac{J_0(k_M r) - J_0(k_M a)}{J_0(k_M a)}\right). \tag{5.3.1}$$

Here, $\Im$ is the radial tension per unit length at the circumference, a is the radius, $k_M = \omega/u_M$, u_M is the speed of transverse waves on the membrane given by

$$u_M = (\Im/\rho_S)^{1/2} \tag{5.3.2}$$

and ρ_S is the mass per unit area. The characteristic shapes of the displaced membrane are thus determined by the zero-order Bessel function $J_0(k_M r)$ and, for low frequencies, equation (5.3.1) approaches the simple form valid for static displacements:

$$\xi(r) = (p_a/4\Im)(a^2 - r^2). \tag{5.3.3}$$

The response of the membrane is also frequency independent in the regime where $k_P a$ is small. This regime may be extended by increasing the tension but the sensitivity is thereby reduced. At higher frequencies the response is punctuated by sharp resonances which occur whenever $k_M a$ is equal to one of the zeros of $J_0(z)$. Denoting these zeros by ψ_{0n}, the resonance frequencies are given by

$$f_n = \psi_{0n}(u_M/2\pi a)$$

$$\psi_{01} = 0.766\pi \quad \psi_{02} = 1.757\pi \quad \psi_{03} = 2.755\pi \quad \psi_{0n} \xrightarrow[n \to \infty]{} (n - \tfrac{1}{4})\pi \tag{5.3.4}$$

Usually we are interested only in the mean surface displacement amplitude of the membrane which is given by

$$\langle \xi \rangle = (p_a/k_M^2 \Im)[J_2(k_M a)/J_0(k_M a)]$$

$$\xrightarrow[\omega \to 0]{} (p_a a^2/8\Im). \tag{5.3.5}$$

$\langle \xi \rangle$ is illustrated in figure 5.5 as a function of frequency for frequencies up to around four times the fundamental. Note that the low-frequency sensitivity is not affected by the thickness of the membrane. Consequently, the mass of the moving element can be made small by use of very thin membranes thereby achieving a high fundamental resonance frequency and a broad bandwidth of uniform response. The limitation in this respect is imposed by the tensile strength of the material and/or by creep under tension. If the maximum stress to which the material may be subjected is $\mathcal{T}_{max}$ then the maximum resonance frequency that can be obtained is $(\chi_1/2\pi a)(\mathcal{T}_{max}/\rho)^{1/2}$, where ρ is the mass density of the membrane material. For stainless steel with a yield stress of order 250 MPa, the maximum fundamental resonance frequency for a 10 mm diameter membrane would be around 14 kHz and the corresponding displacement sensitivity, $\langle\xi\rangle/p_a$, about 2.5 nm Pa^{-1}. Capacitance microphones with tensioned nickel diaphragms are available commercially with resonance frequencies up to 100 kHz [4]. One method by which both high tension and good long-term mechanical stability can be achieved consists of stretching the membrane to a tension close to the elastic limit and then subjecting it to a process of artificial ageing at an elevated temperature during which some relaxation takes place.

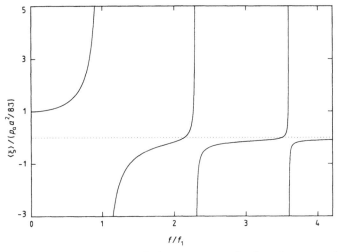

Figure 5.5 Mean surface displacement amplitude of a circular membrane driven by acoustic pressure p_a acting on one side.

When operated in gases, the loading of a vibrating membrane by the surrounding fluid is usually slight and the sensitivity differs little from that illustrated except near to a resonance where the viscosity of the fluid provides damping. It is possible to arrange for the damping of the fundamental resonance to be sufficient to smooth out the response so that a nearly uniform sensitivity is achieved for frequencies up to a little above f_1. In dense gases or liquids, the mass of the fluid around the membrane must be taken into account; this serves primarily to lower the frequency of the fundamental resonance. Quite sophisticated calculations of the effects of the surrounding fluid on sensitivity are possible for transducers whose vibrating elements are tensioned membranes (or stiff plates); the interested reader is referred to the literature [5, 6]. For our present purposes equation (5.3.5) is usually sufficient; it will allow estimation of transducer sensitivity, frequency response and mechanical impedance. Approximate methods are available by which the loading effects of the fluid and of mechanical linkages or additional masses may be included (see §5.3.3).

5.3.2   The circular plate

Unlike the membrane, which derives its stiffness through an applied radial tension, plates have intrinsic stiffness and are not usually subjected to tension. However, there is now a choice of boundary conditions at the circumference including edge-pinned and edge-clamped plates; we shall consider only the latter and restrict our attention to circularly symmetric vibrations. For such an undamped plate of thickness d and radius a driven on one side by a uniform acoustic pressure $p_a \exp(i\omega t)$ the displacement amplitude is [7]

$$\xi(r) = \left(\frac{I_1(k_P a) J_0(k_P r) + J_1(k_P a) I_0(k_P r)}{J_0(k_P a) I_1(k_P a) + I_0(k_P a) J_1(k_P a)} - 1 \right) \left(\frac{p_a}{\rho \omega^2 d} \right) \quad (5.3.6)$$

where

$$k_P^4 = [12\omega^2 \rho (1 - \sigma^2) / Y d^2]. \quad (5.3.7)$$

In these equations, the plate material is characterized by mass density ρ, Poisson's ratio σ and Young's modulus Y. The functions $I_m(z)$ are hyperbolic (or modified) Bessel functions and are given by $I_m(z) = i^{-m} J_m(iz)$. At low frequencies, equation (5.3.6) reduces to the form appropriate to static deflections:[2]

$$\xi(r) = [3p_a(1 - \sigma^2) / 16 Y d^3] (a^2 - r^2)^2. \quad (5.3.8)$$

We see that the characteristic shape of the deflection differs from that for the tensioned membrane even in the static limit. This difference in shape arises mainly because of the difference in boundary conditions which in the present case require not only ξ but also $\partial \xi / \partial r$ to vanish on $r = a$. The resonance frequencies of the plate correspond to the values of $k_P a$ for which the denominator of equation (5.3.6) vanishes and are given by

$$f_n = \nu_n^2 (d/4\pi a^2) [Y/3\rho(1 - \sigma^2)]^{1/2}$$

$$\nu_1 = 1.015\pi \quad \nu_2 = 2.007\pi \quad \nu_3 = 3.000\pi \quad \nu_n \xrightarrow[n \to \infty]{} n\pi. \quad (5.3.9)$$

The mean surface displacement amplitude of the plate is

$$\langle \xi \rangle = \left(\frac{4(k_P a)^{-1} J_1(k_P a) I_1(k_P a)}{J_0(k_P a) I_1(k_P a) + I_0(k_P a) J_1(k_P a)} - 1 \right) \left(\frac{p_a}{\rho \omega^2 d} \right)$$

$$\xrightarrow[\omega \to 0]{} [(1 - \sigma^2) a^4 / 16 Y d^3] p_a. \quad (5.3.10)$$

This is illustrated in figure 5.6 as a function of frequency for frequencies

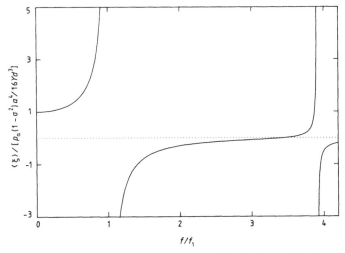

Figure 5.6 Mean surface displacement amplitude of an edge-clamped circular plate driven by acoustic pressure p_a acting on one side.

up to around four times the fundamental. Since the thickness controls both the low-frequency sensitivity and the fundamental resonance frequency, these quantities are not easily modified independently. As an example of typical plate parameters, consider a stainless-steel plate ($Y \approx 200$ GPa, $\sigma \approx 0.27$) 0.1 mm thick and 10 mm in diameter. This has a fundamental frequency of about 10 kHz and a low-frequency displacement sensitivity of about 0.2 nm Pa^{-1}. As with the stretched membrane, one can trade low-frequency broad-band sensitivity ($\sim a^4/d^3$) for increased resonance frequency ($\sim d/a^2$). Of course, narrow-bandwidth applications are also possible and plates are particularly suitable for applications at frequencies below about 100 kHz where conventional piezoelectric transducers are too large for many applications.

5.3.3 Lumped circuit analysis

Many applications of vibrating diaphragms in transducer systems involve operation at frequencies only up to about the first resonance. Under these conditions, a simple and useful model of diaphragm motion is afforded by the lumped circuit analysis. This model has the advantage that the effects of transducer loading may be incorporated, using additional circuit elements, without incurring the penalty of greatly increased mathematical complexity. Since practical systems are often less than perfectly characterized, an approximate treatment is usually all that is required.

The mean displacement amplitude is written in terms of the mechanical impedance Z_m:

$$\langle \xi \rangle = p_a \pi a^2 / i\omega Z_m \qquad (5.3.11)$$

which we approximate with the series combination of a damping resistance R_m, a stiffness s, and an effective inertial mass m_{eff},

$$Z_m \approx R_m + i(\omega m_{eff} - s/\omega) \qquad (5.3.12)$$

so that the displacement amplitude becomes

$$\langle \xi \rangle = p_a \pi a^2 / (s - \omega^2 m_{eff} + i\omega R_m). \qquad (5.3.13)$$

The zero-frequency sensitivity is determined by the stiffness and the resonance frequency by s/m_{eff}. The damping resistance is important only near to resonance where it determines the quality factor $Q = \omega_1 m_{eff}/R_m$. Note that the model predicts only one resonance. The stiffness is given in terms of the ratio of a uniform static force $\pi a^2 p$ and the displacement that results from that force:

$$s = (\pi a^2 p / \langle \xi \rangle)_{\omega \to 0}. \qquad (5.3.14)$$

This may be calculated from the classical theory of elasticity without consideration of the dynamic displacement. The effective mass of the system is chosen with the criterion that the kinetic energy of the model system, in which a lumped mass m_{eff} undergoes motion with displacement amplitude $\langle \xi \rangle$, equals that of the real system. Since the velocity amplitude is proportional to the displacement amplitude for harmonic motion at fixed frequency, the most general requirement of this criterion may be written

$$\tfrac{1}{2} m_{eff} \langle \xi \rangle^2 = \frac{1}{2} \int_0^a \int_0^{2\pi} \rho_S(r, \theta) [\xi(r)]^2 \, d\theta \, r dr \qquad (5.3.15)$$

where the mass per unit area is given as a function of the polar coordinates (r, θ). In order to arrive at a frequency-independent effective mass, equation (5.3.15) is evaluated using the expressions for $\langle \xi \rangle$ and $\xi(r)$ appropriate to static displacements.

In the case of the uniform tensioned membrane we have stiffness $8\pi\mathfrak{I}$ and effective mass $(4/3)\rho_S \pi a^2$. The low-frequency sensitivity is given exactly by the model and the predicted resonance frequency is only about 1.8 per cent above that obtained by the rigorous treatment. Similarly, the edge-clamped plate may be characterized by a stiffness $16\pi Y d^3/(1 - \sigma^2)a^2$ and an effective mass $(9/5)d\rho\pi a^2$; the predicted resonance frequency then exceeds that of equation (5.3.9) by just 1.6 per cent. In both cases, the model differs little at frequencies below resonance from the more complicated expressions given above. It may be useful to compare the properties of membranes and plates as diaphragms for wide-band transducers. The most meaningful comparison here is between two diaphragms that have the same fundamental resonance frequency and therefore the same useful bandwidth. A membrane and a plate made from the same material will satisfy this condition when the tension of the former is approximately equal to $40Yhd^2 a_M^2/27(1 - \sigma^2)a_P^4$, where h is the thickness of the membrane, and a_M and a_P are the radii of the membrane and the plate respectively. The ratio s_P/s_M of plate (s_P) and membrane (s_M) stiffness is then $27 da_P^2/20ha_M^2$. Typical membrane thicknesses of 5 to 10 μm compare with typical plate thicknesses of 25 to 100 μm and indicate that membranes usually offer much greater sensitivity than a comparable plate for the same bandwidth and diameter. However, plate diaphragms would appear to be advantageous when very small diameters are contemplated.

Coupling between the motion of the diaphragm and that of other parts of a real system may be incorporated by inclusion of additional circuit elements. For example, the radiation of sound into the surrounding fluid may be included by taking the series combination of the radiation impedance and the impedance of the unloaded diaphragm. The effect of direct mass loading (e.g. the coil of a moving-coil transducer) may be incorporated through equation (5.3.15) by using an appropriate function for the mass per unit area.

5.3.4   Non-linear behaviour

For transducers driven under non-resonant conditions, typical displacement amplitudes are extremely small and the motion should be strictly linear even when the damping is small. However, if the transducer is driven at one of its mechanical resonance frequencies then the displacement is limited only by the damping and, when driving a gaseous medium, can become quite large. Under these conditions non-linearity can be appreciable and may, under certain circumstances, become a source of systematic error in acoustic interferometry [8].

A simple description of the main kinds of non-linear behaviour can be derived by considering a one-dimensional oscillator consisting of a mass m moving against a mechanical resistance R_m under the influence of a harmonic driving force and a non-linear restoring force. Since, for non-linear problems, real and imaginary terms in the equation of motion do not separate, it is not possible to continue to use the time dependence $\exp(i\omega t)$; instead, we write the driving force as $F\cos(\omega t)$. The equation of motion may then be written in the form

$$F\cos(\omega t) = R_m \frac{dx}{dt} + m \frac{d^2 x}{dt^2} + s_0 x + s_1 f(x) \qquad (5.3.16)$$

where x is the displacement, $s_0 x$ is the linear restoring force, and $s_1 f(x)$ is the non-linear term controlled by a function of displacement $f(x)$ and a parameter s_1. Most kinds of non-linear behaviour may be described by either the symmetric perturbation function $f(x) = x^2$ or the asymmetric function $f(x) = x^3$. In either case, steady-state periodic solutions of the equation of motion may be developed as a Fourier series with fundamental angular frequency ω:

$$x(t) = \sum_n a_n \cos(n\omega t + \delta_n). \qquad (5.3.17)$$

Here a_n and δ_n (both real) are the amplitude and phase of the nth term in the series.[3] The general behaviour is well known. For the symmetric perturbation only even terms in the series have non-vanishing amplitudes while for asymmetrically perturbed oscillators only odd terms appear. Usually some kind of lock-in detection is used that rejects frequencies other than the fundamental and our interest will then lie with the $n = 1$ component of the series which is affected only by asymmetric non-linear perturbations. To simplify the notation, the equation of motion for this case is written in the dimensionless form

$$\frac{d^2\xi}{d\theta^2} + \frac{1}{Q}\frac{d\xi}{d\theta} + \xi(1 + \varepsilon\xi^2) = \cos(\gamma\theta) \qquad (5.3.18)$$

where

$$\xi = s_0 x / F \qquad\qquad \theta = \omega_0 t = (s_0/m)^{1/2} t$$
$$\varepsilon = s_1 F^2 / s_0^3 \qquad\qquad Q = m\omega_0 / R_m = s_0 / \omega_0 R_m$$
$$\gamma = \omega/\omega_0.$$

In this reduction, the dimensionless displacement amplitude ξ is obtained by dividing x by the static displacement F/s_0 that would be obtained in the absence of non-linear behaviour. Thus for small driving forces $\xi \to 1$ at low frequencies. The dimensionless variables corresponding to time and frequency, θ and γ, are defined in terms of the angular resonance frequency ω_0 found in the linear regime. The parameter ε measures the magnitude of the non-linear restoring force. Positive values of ε give a 'hard' restoring force (effective stiffness increases with displacement) while negative values of ε correspond to a 'soft' restoring force. Since at resonance ξ is of order Q, the condition for non-linear effects to be small is $|\varepsilon|Q^2 \ll 1$.

The method used here to solve the equation of motion is that described by Morse and Ingard [9]. ξ is expanded as a Fourier series with fundamental frequency equal to the driving frequency and attention is restricted to the leading term. Thus we have

$$\xi = \xi_1 \cos(\gamma\theta + \delta_1) \qquad (5.3.19)$$

which, when substituted into equation (5.3.18), yields

$$A[\cos(\delta_1) - \tan(\gamma\theta)\sin(\delta_1)] + B[\sin(\delta_1) + \tan(\gamma\theta)\cos(\delta_1)] = 1 \quad (5.3.20)$$

where

$$A = [(1 - \gamma^2) + \tfrac{3}{4}\varepsilon\xi_1^2]\xi_1 \qquad B = -(\gamma/Q)\xi_1.$$

In order that this equation be satisfied, the amplitude must be a solution of

$$A^2 + B^2 = (\gamma/Q)^2 \xi_1^2 + [(1 - \gamma^2) + \tfrac{3}{4}\varepsilon\xi_1^2]^2 \xi_1^2 = 1 \qquad (5.3.21)$$

and the phase must be given by

$$\tan(\delta_1) = B/A = -\frac{\gamma/Q}{(1 - \gamma^2) + \tfrac{3}{4}\varepsilon\xi_1^2}. \qquad (5.3.22)$$

In general, the main effect of non-linearity on the fundamental term is to introduce a shift in the 'resonance' frequency of the oscillator initially proportional to the square of the driving force. One can show that for a slightly non-linear oscillator with a high Q the fractional change in the resonance frequency is $\tfrac{3}{8}\varepsilon Q^2$. Thus, a transducer driven at its linear resonance frequency may be forced off resonance as the amplitude is increased. This kind of behaviour is illustrated in figure 5.7 for an oscillator with a quality factor $Q = 50$. The linear case and two levels of non-linearity are shown here. When εQ^2 is small (0.02 in the example illustrated) the oscillator remains stable for driving frequencies near its linear resonance frequency and there is just a small shift in the frequency of maximum response. However, for larger values of εQ^2 the non-linear term has catastrophic effects; the response function is multivalued over part of the frequency range and the amplitude can jump suddenly from one value to another. When $\varepsilon > 0$ (the cases illustrated) this behaviour occurs above, and for $\varepsilon < 0$ below, the linear resonance frequency.

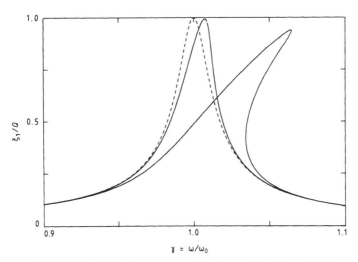

Figure 5.7 Reduced displacement amplitude as a function of reduced frequency for linear (– – – –) and non-linear (————) oscillators. Two degrees of non-linearity are shown: εQ^2 equal to 0.02 and 0.20.

Since the speed is $-(\omega F/s_0)\xi_1 \sin(\omega t + \delta_1) = (\omega F/s_0)\xi_1 \cos(\omega t + \delta_1 \pi/2)$, the effective mechanical impedance Z_{eff} of the oscillator at the driving frequency has magnitude $s_0/\omega\xi_1$ and phase $\phi = -(\delta_1 + \pi/2)$. Returning to complex notation Z_{eff} may be written

$$Z_{\text{eff}} = (s_0/\omega\xi_1)\exp(i\phi) \qquad (5.3.23)$$

where the phase $\phi = -(\delta_1 + \pi/2)$. In practice the mechanical impedance in question will be the sum of the transducer's intrinsic impedance Z_T and the impedance Z_L of the load presented to it by the fluid. Non-linearity is most likely to cause problems in constant-frequency variable pathlength interferometry because the transducer is usually driven at one of its mechanical resonance frequencies. However, since the oscillator in the model represents not just the transducer but the combined system of transducer and acoustic load, the dimensionless frequency γ is not always exactly unity. Assuming that the frequency is constant and equal to the linear resonance frequency of the unloaded transducer, the condition $\gamma = 1$ is met exactly only when the cavity is brought into resonance (where $\text{Im}(Z_L) = 0$).

Provided that the non-linear effects are small an approximate solution correct to first order in ε is sufficient. Since in zeroth order we have $\xi_1 = [(\gamma/Q)^2 + (1 - \gamma^2)^2]^{-1/2}$ the non-linear term in equation (5.3.21) may be estimated and first-order expressions for the reduced amplitude and the phase angle obtained:

$$\xi_1^{-1} \approx [(\gamma/Q)^2 + (1 - \gamma^2)^2]^{1/2}\left(1 - \frac{3}{4}\varepsilon\frac{(1 - \gamma^2)}{[(\gamma/Q)^2 + (1 - \gamma^2)^2]^2}\right) \quad (5.3.24)$$

and

$$\tan(\phi) = A/B \approx (\gamma^2 - 1)(Q/\gamma) - \frac{3}{4}\varepsilon\left(\frac{Q/\gamma}{[(\gamma/Q)^2 + (1 - \gamma^2)^2]}\right). \quad (5.3.25)$$

At the linear resonance frequency, the magnitude of Z_{eff} is unperturbed in leading order but the phase angle ϕ is rotated from zero to $\phi_1 = -\frac{3}{4}\varepsilon Q^3$; zero phase is reached at the perturbed 'resonance' frequency $\gamma = 1 + \frac{3}{8}\varepsilon Q^2$. This can lead to errors in acoustic interferometry when absolute phase information is used.

5.4 The Piezoelectric Effect

The piezoelectric effect, by which electric polarization is generated within a solid material when it is subjected to a mechanical stress, is exhibited by certain materials that lack inversion symmetry. If a piezoelectric crystal is subjected to a mechanical stress then the resulting deformation may cause an electric field to be created within it and positive and negative charges to appear on opposite faces. Conversely, when subjected to an externally applied electric field, the material may undergo a deformation. The magnitude of the effect depends upon the symmetry of the crystal and its orientation relative to the external field.

Piezoelectric materials may be used to generate or detect sound fields in several ways. Thin plates or cylinders of the material may be placed in direct contact with the fluid so that longitudinal waves in the fluid are coupled with compression waves in the transducer. In order to obtain useful signal levels, operation at a resonance frequency of the piezoelectric element is usually required. Alternatively, thin discs or strips of piezo-electric material may be coupled to the motion of a stiff diaphragm and operated in a bending mode. Efficient broad-band transducers of this kind can be designed.

5.4.1 Piezoelectric materials

Although a very large number of materials exhibit piezoelectric behaviour in some degree, only a relatively modest number have a combination of electrical and mechanical properties suitable for their use in the generation and detection of sound pressure waves. However, the class of useful piezoelectric materials does include a number of polycrystalline and amorphous solids as well as single crystals.

Quartz, which is naturally piezoelectric and can be found as large speci-mens, is the most common example of a single-crystal piezoelectric material with useful properties as a transducer. Figure 5.8 depicts a crystal of α quartz; this is the piezoelectric phase and is stable under atmospheric pressure at temperatures below 846 K. The principal axis of symmetry is the threefold (C_3) axis which we take as the z axis of a Cartesian coordinate system. Perpendicular to the principal axis are three twofold (C_2) axes each connecting two opposite edges of a six-sided prism. One of these C_2 axes is taken as the x axis, leaving the direction perpendicular to both x and z to be labelled as the y axis.[4] It is simplest to consider a block of material cut from the crystal with its edges parallel with these Cartesian axes. As shown in figure 5.9(a), the faces perpendicular to the x axis are labelled by 1, those perpendicular to y by 2, and those perpendicular to z by 3. In quartz, no combination of stresses can produce charge on faces '3'. Tensile stresses always produce charge on the '1' faces, while shear stresses can produce charge on either the '1' or the '2' faces. Practical transducers are usually fabricated from thin slices of material cut at various angles with respect to the crystallographic axes. Electrodes for imposing or detecting the electric field may be deposited on the faces of the transducer by plating or by vacuum deposition. The most common orientation for use in the detection of longitudinal sound waves is the x-cut slice (large faces '1') which, as shown in figure 5.9(b), functions as an expander plate. There is also coupling between the x component of the electric field and strain along the y axis (figure 5.9(c)). Other important orientations include y-cut plates (figure 5.9(d)), and various rotated y cuts, that operate in the thickness shear mode. Because of the high mechanical stability and low internal damping in quartz, this material is an excellent choice for narrow-bandwidth applications with the crystal acting directly on the fluid and operating at one of its natural resonance frequencies. Non-resonant operation is usually too inefficient. Quartz has a relatively small dielectric constant but this and the other parameters characterizing its piezoelectric properties are quite stable over a wide temperature range.

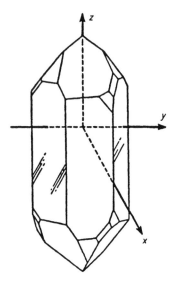

Figure 5.8 A crystal of α quartz showing the coordinate system used in the text.

Most of the other piezoelectric materials in modern use are either ceramics or polymers. Piezoelectric ceramics are fabricated from polycrystalline ferroelectric materials such as lead zirconate titanate[5] and barium titanate. Consequently, below the upper ferroelectric critical temperature (Curie temperature) the individual crystallites that comprise the ceramic possess a spontaneous electric polarization which may be modulated by applied stress. However, because of the initially random orientation of crystallites in the ceramic mix, these materials must be subjected to a process of poling to obtain piezoelectric properties in the bulk sample. This is achieved by cooling the sample in a large static electric field from a temperature above the Curie point, thereby imposing a preferred orientation parallel to the field. The properties of the ceramic remain isotropic in the plane perpendicular to the poling axis. Piezoelectric ceramics are characterized by high dielectric constants and much larger electro-mechanical coupling than found in quartz. However, their properties vary quite strongly with temperature and the maximum operating temperatures are lower.

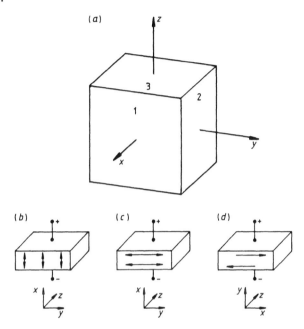

Figure 5.9   Principal cuts of quartz: (*a*) coordinate system used in the text; (*b*) *x*-cut plate in thickness mode; (*c*) *x*-cut plate in longitudinal mode; (*d*) *y*-cut plate in thickness shear mode.

Piezoelectric polymers, such as polyvinylidene fluoride (PVDF), are usually manufactured in the form of thin films that may be polarized in a direction perpendicular to the plane by a number of techniques. In addition, the polymer film may be stretched during manufacture to estab-lish the molecular orientation in two dimensions. Electrodes are usually formed by vacuum deposition of aluminium or gold. Piezoelectric poly-mers have dielectric constants not far in excess of those of other non-polarized polymers and fairly low electromechanical coupling constants. However, they are much more compliant than crystals or ceramics and gen-erate large EMFs under applied stress. Thermal depoling of the polymer generally restricts the maximum operating temperature to considerably below that for non-piezoelectric applications of the same material.

For ceramics and polymers, the poling axis is conventionally taken as the *z* axis. Ceramic plates cut perpendicular to the poling axis may be operated in either the thickness or length expander modes, while those cut parallel to the poling axis are suitable for operation in the thickness shear mode; similar orientations are possible for piezoelectric polymer films. Both these kinds of material have much greater internal damping than single-crystalline materials and therefore offer enhanced bandwidth when operated near resonance. This is of great advantage in pulsed operation.

5.4.2   Theory of piezoelectricity

In an elastic solid, the stress $\mathscr{T}$ and strain $\mathscr{S}$ may be represented by symmetric second-rank tensors. The stress may be written in terms of the six unique elements of the tensor $\mathscr{T}$:

$$
\begin{aligned}
\mathscr{T}_1 &= \mathscr{T}_{xx} & \mathscr{T}_4 &= \mathscr{T}_{yz} = \mathscr{T}_{zy} \\
\mathscr{T}_2 &= \mathscr{T}_{yy} & \mathscr{T}_5 &= \mathscr{T}_{xz} = \mathscr{T}_{zx} \\
\mathscr{T}_3 &= \mathscr{T}_{zz} & \mathscr{T}_6 &= \mathscr{T}_{xy} = \mathscr{T}_{yx}
\end{aligned}
\tag{5.4.1}
$$

where $\mathscr{T}_1$ to $\mathscr{T}_3$ represent the tensile stresses acting along the three Cartesian axes and $\mathscr{T}_4$ to $\mathscr{T}_6$ represent the three possible shear stresses. The strain is defined in terms of relative deformation and, denoting displacement along the x, y, z axes by ξ_1, ξ_2, ξ_3, the six unique components of the strain tensor may be written

$$
\begin{aligned}
\mathscr{S}_1 &= (\partial \xi_1/\partial x) & \mathscr{S}_4 &= (\partial \xi_2/\partial z) + (\partial \xi_3/\partial y) \\
\mathscr{S}_2 &= (\partial \xi_2/\partial y) & \mathscr{S}_5 &= (\partial \xi_3/\partial x) + (\partial \xi_1/\partial z) \\
\mathscr{S}_3 &= (\partial \xi_3/\partial z) & \mathscr{S}_6 &= (\partial \xi_1/\partial y) + (\partial \xi_2/\partial x)
\end{aligned}
\tag{5.4.2}
$$

where $\mathscr{S}_1$ to $\mathscr{S}_3$ are tensile strains and $\mathscr{S}_4$ to $\mathscr{S}_6$ are shear strains. It is convenient to adopt a matrix notation in which the six unique elements of each tensor are arranged into a 6×1 matrix (six rows, one column).

In a non-piezoelectric material, the relation between stress and strain is, for small deformations, a linear one determined by a set of 36 elastic constants c_{jk}. In matrix notation, this relation may be written

$$
\mathscr{T} = c\mathscr{S}
\tag{5.4.3}
$$

where c is the 6×6 matrix with elements c_{jk} formed from the elastic moduli. The inverse of this relation,

$$
\mathscr{S} = s\mathscr{T}
\tag{5.4.4}
$$

involves the compliance matrix $s = c^{-1}$.[6]

Similarly, in a common dielectric, the relation between the electric flux density D and the electric field E (both vectors) is also linear,

$$
D = \varepsilon E
\tag{5.4.5}
$$

and involves a set of nine dielectric permittivities represented here by the

3×3 matrix $\boldsymbol{\varepsilon}$. The inverse of this relation,

$$E = \beta D \qquad (5.4.6)$$

is given in terms of the dielectric impermeability matrix: $\beta = \boldsymbol{\varepsilon}^{-1}$.

For a piezoelectric material, the elastic and electrical properties are coupled and these relations must be modified. Four different representations are possible depending on which combination of one tensor and one vector is chosen for the independent variables. With $\mathscr{S}$ and E as independent variables, the generalized expressions for stress and electric flux density become

$$\mathscr{T} = c^E \mathscr{S} - e_t E \qquad (5.4.7a)$$

$$D = e \mathscr{S} + \boldsymbol{\varepsilon}^S E \qquad (5.4.7b)$$

Here e is a 3×6 matrix of piezoelectric stress coefficients and e_t denotes its transpose. Superscripts indicate that the elastic constants which appear here are defined for conditions of constant or zero electric field, while the dielectric constants are defined for conditions of constant or zero strain. When $\mathscr{T}$ and E are taken as independent variables, the corresponding constitutive relations may be written

$$\mathscr{S} = s^E \mathscr{T} - d_t E \qquad (5.4.8a)$$

$$D = d \mathscr{T} + \boldsymbol{\varepsilon}^T E \qquad (5.4.8b)$$

where d is a 3×6 matrix of piezoelectric strain coefficients. The dielectric constants $\boldsymbol{\varepsilon}^T$ refer to conditions of constant stress and may differ considerably from the constant strain ('clamped') values $\boldsymbol{\varepsilon}^S$.

Using elastic constants defined for conditions of constant D, one may select as independent variable either $\mathscr{S}$ and D

$$\mathscr{T} = c^D \mathscr{S} - h_t D \qquad (5.4.9a)$$

$$E = -h \mathscr{S} + \beta^S D \qquad (5.4.9b)$$

or, finally, $\mathscr{T}$ and D

$$\mathscr{S} = s^D \mathscr{T} - g_t D \qquad (5.4.10a)$$

$$E = -g \mathscr{T} + \beta^T D. \qquad (5.4.10b)$$

In these representations, the coefficients of the h matrix relate electric field and strain while those of the g matrix relate electric field and stress; both are defined under open-circuit (constant D) conditions.

One measure of the magnitude of the piezoelectric effect is provided by the electromechanical coupling coefficient K. K^2 is defined as the fraction of the stored dielectric energy available to do mechanical work (also equal to the fraction of stored elastic energy available to do electrical work) and depends upon the kind of deformation in question.

On grounds of symmetry, the number of unique constants in a given representation is usually very much less than the 36 elastic terms, 18 piezoelectric terms and nine dielectric terms that appear in these equations.[7] In quartz, for example, there are, in a given representation, 18 non-zero elastic constants of which just six are unique, five non-zero piezoelectric coefficients of which only two are unique and three non-zero dielectric constants of which two are unique. The sets of piezoelectric constants appearing in the different representations are not independent; indeed they are interrelated by linear matrix equations involving either dielectric constants or elastic constants. Also, in most materials including quartz and the piezoelectric ceramics, the off-diagonal elements in both the permittivity and impermeability matrices vanish so that the diagonal elements of the two matrices are related by $\beta_{kk} = 1/\varepsilon_{kk}$. Finally, when a particular symmetry of motion is considered it is often possible to reduce the constitutive relations to a one-dimensional representation involving only scalar quantities.

A few of the properties of representative piezoelectric materials are given in table 5.2. The piezoelectric coefficients are given only for coupling of parallel electric and elastic fields: 1, 1 coefficients for quartz and 3, 3 coefficients for the other materials. The maximum operating temperature $T_{\max}$ given in the table is the α–β transition temperature for quartz, the Curie temperature for the ceramics PZT and barium titanate and an approximate temperature for thermal depoling of PVDF.

Table 5.2   Selected properties of piezoelectric materials at 300 K [10–12].

	Quartz	PZT-4	PZT-5	BaTiO$_3$	PVDF
			Fields parallel		
$d/(10^{-12}\,C\,N^{-1})$	2.3	289	374	190	-33
$e/(C\,m^{-2})$	0.17	15.1	15.8	17.5	—
$g/(10^{-3}\,m^2\,N^{-1})$	50	26.1	24.8	12.6	-339
$h/(10^8\,V\,m^{-1})$	43	26.8	21.5	15.6	—
$\varepsilon^S/\varepsilon_0$	4.48	635	830	1260	6
$\varepsilon^T/\varepsilon_0$	4.52	1300	1700	1700	12
K_t	0.095	0.51	0.49	0.38	0.15
c^E/GPa	86	115	111	146	8.9
c^D/GPa	87	159	147	171	9.1
s^E/TPa^{-1}	12.8	15.5	18.8	9.5	—
s^D/TPa^{-1}	12.7	7.90	9.46	7.1	—
$\rho/(kg\,m^{-3})$	2650	7500	7750	5700	1780
$u^D/(m\,s^{-1})$	5730	4600	4360	5480	2200
$T_{\max}/K$	846	601	638	388	≈ 370

$\varepsilon_0 = 8.854\,pF\,m^{-1}$.

5.4.3   Thickness expanders

The most common type of piezoelectric transducer for direct action on the fluid is fabricated in the form of a thin plate with electrodes on both major surfaces and operated at resonance in the thickness expander mode. In view of the importance of this kind of device, it is worth while analysing its mechanical and electrical properties in some detail. We consider a piezoelectric plate of surface area $\mathcal{C}$ and thickness $2l$ mounted perpendicular to the x axis such that the plane $x = 0$ lies midway between the two faces. Since the plate is assumed thin, the mechanical boundary conditions approach those of lateral clamping and we have a condition of one-dimensional strain: $S_1 \neq 0$, $S_{j > 1} = 0$. The major faces of the transducer, being plated, are equipotential surfaces; so, in the absence of significant dielectric leakage, the electrical boundary conditions require that $D_2 = D_3 = 0$. Furthermore, since piezoelectrics are insulating materials containing no free charges, Gauss' law requires that the remaining component of the flux density, D_1, be independent of x and equal to the surface charge density on the electrodes. Consequently, when all fields have a simple harmonic variation in time, $D_1 = I/i\omega\mathcal{C}$ (where I is the electric current). In view of these boundary conditions, the natural choice for the independent variables is that of the $(\mathcal{S}, D)$ representation (equation (5.4.9)) in terms of which the stress and electric field across the plate are given by

$$\mathcal{T}_1 = c_{11}^D \mathcal{S}_1 - h_{11} D_1 \qquad (5.4.11)$$

$$E_1 - h_{11} \mathcal{S}_1 + \beta_{11}^S D_1. \qquad (5.4.12)$$

The coefficients here are those appropriate for x-cut quartz. However, any of the formulae in this section can be applied to ceramics or polymers by a change of notation from 1, 1 components to 3, 3 components.

Writing the strain $\mathcal{S}_1$ as $(\partial \xi/\partial x)$, where ξ is the displacement, a wave equation for the plate is obtained directly from equation (5.4.11) by applying Newton's second law, which requires $(\partial \mathcal{T}_1/\partial x) = \rho(\partial^2 \xi/\partial t^2)$, and recalling that $(\partial D_1/\partial x) = 0$:

$$(\partial^2 \xi/\partial x^2) + (\rho/c_{11}^D)^2 \xi = 0. \qquad (5.4.13)$$

The phase speed for longitudinal sound waves u^D in the crystal is therefore given in terms of the constant D modulus of elasticity by

$$u^D = (c_{11}^D/\rho)^{1/2}. \qquad (5.4.14)$$

The constant-flux-density elastic modulus that appears here exceeds the constant-field elastic modulus because of the stiffening effect of the electric field created when the material is compressed under open-circuit conditions. In fact, from equations (5.4.11) and (5.4.12), c_{11}^D and

$c_{11}^E = (\partial \mathcal{T}_1/\partial \mathcal{S}_1)_E$ are related by

$$(c_{11}^D - c_{11}^E)/c_{11}^E = K_t^2 \tag{5.4.15}$$

where

$$K_t = (h_{11}^2/\beta_{11}^S c_{11}^D)^{1/2} \tag{5.4.16}$$

is the electromechanical coupling constant for this mode of motion and $\beta_{11}^S = 1/\varepsilon_{11}^S$. $K_t \approx 0.1$ for quartz but piezoelectric ceramics are characterized by much greater values and the constant-flux-density sound speed may exceed the constant-field value by 20 or 30 per cent.[8]

Equations (5.4.11–13) are the working equations of the plate transducer; now that they have been established, the discussion will proceed as follows. First, the input mechanical admittance presented by one face of the transducer is calculated and used to obtain the mechanical resonance frequencies in vacuum as a function of the electrical load impedance. Next, the effect of the fluid loading is included and the operation of the device as a detector is described. Finally, the electrical input admittance is calculated as a function of the acoustical load admittance and the behaviour of the transducer as a source of sound is described.

To calculate the input mechanical admittance we assume that the positive x side of the transducer is subject to a uniform applied stress $(F/\mathcal{Q})\exp(i\omega t)$ and that the resulting speed is $v \exp(i\omega t)$, both being positive in the negative x direction. For the moment, the other face of the transducer is assumed to be free so that the mechanical boundary condition there is $\mathcal{T}_1(x = -l) = 0$. An appropriate solution of the wave equation for such asymmetric boundary conditions is

$$\xi(x, t) = [\xi_0 \sin(kx) + \xi_1 \cos(kx)] \exp(i\omega t) \tag{5.4.17}$$

where $k = \omega/u^D$. However, before attempting to evaluate the two constants ξ_0 and ξ_1, we must specify the electrical boundary conditions. To do this, first note that the potential difference generated between the electrodes is given from equation (5.4.12) by

$$V = - \int_{-l}^{l} E_1 \, dx = h_{11}\Delta\xi - 2l\beta_{11}^S D_1 \tag{5.4.18}$$

where $\Delta\xi = 2\xi_0 \sin(kl)$ is the relative displacement of the two faces. Then, if the electrical impedance of the external circuit is denoted by Z_e, so that $V = i\omega\mathcal{Q}D_1 Z_e$, then the electric flux density between the plates becomes

$$D_1 = (h_{11}\Delta\xi/2l\beta_{11}^S)(1 + i\omega C_0 Z_e)^{-1} \tag{5.4.19}$$

where

$$C_0 = \mathcal{Q}/2l\beta_{11}^S \tag{5.4.20}$$

is the clamped (constant strain) capacitance of the transducer. Equations (5.4.17) and (5.4.19) may now be used to express both strain and flux density in terms of ξ_0 and ξ_1, which can then be obtained by solving (5.4.11) subject to the mechanical boundary conditions. This leads to the following expression for the input mechanical admittance:

$$Y_m = v/F = -(\mathrm{i}\omega/F)[\xi_0 \sin(kl) + \xi_1 \cos(kl)]$$

$$= \frac{\mathrm{i}/2\alpha\rho u}{\cot(kl) - (K_t^2/kl)(1 + \mathrm{i}\omega C_0 Z_e)^{-1}} - \frac{(\mathrm{i}/2\alpha\rho u)}{\tan(kl)}. \qquad (5.4.21)$$

From this expression we see that, under open-circuit conditions ($Z_e = \infty$), the plate is resonant whenever $kl = n(\pi/2)$ with integer n. Odd values of n correspond to resonance of the first term, while even values correspond to resonance of the second term.

Let us concentrate on the second term in equation (5.4.21) which is associated with the symmetric displacement $\xi_1 \cos(kx)$. At very low frequencies, this term dominates and describes a linear translation of the whole transducer back and forth along the x axis with mechanical admittance $Y_m = 1/\mathrm{i}\omega m$ controlled by the mass m. However, since $\cos(kx)$ is symmetric about $x = 0$, there is no contribution at any frequency to the relative displacement $\Delta\xi$ of the two faces and the even-order resonances of the plate are therefore piezoelectrically inactive. Furthermore, near to the odd-order resonances, the symmetric term gives very small surface displacements and may be neglected altogether without significant inaccuracy.

The first term in equation (5.4.21) is associated with the asymmetric displacement $\xi_0 \sin(kx)$ which determines $\Delta\xi$ and is coupled to the electric field. In general, this term is resonant whenever

$$\mathrm{Re}\,[\cot(kl) - (K_t^2/kl)(1 + \mathrm{i}\omega C_0 Z_e)^{-1}]$$

vanishes and so the frequencies of these resonances depend upon the electrical load impedance. If Z_e is purely resistive and equal to R_L then the resonance frequencies are given by the solutions of

$$(K_t^2/kl)\tan(kl) = 1 + (\omega C_0 R_L)^2. \qquad (5.4.22)$$

Under open-circuit conditions, the resonance frequencies are given by

$$f_n^p = n(u^D/4l) \qquad n = 1, 3, 5, \cdots \qquad (5.4.23)$$

and are known as the parallel resonance frequencies, while the corresponding frequencies under short-circuit conditions are slightly lower and are known as series resonance frequencies f_n^s. When $K_t^2 \ll 1$, as is usually the case, f_n^s is given approximately by

$$f_n^s = n(u^D/4l)[1 - (2K_t/n\pi)^2] \qquad n = 1, 3, 5, \cdots \qquad (5.4.24)$$

so that the fractional difference $(f_n^p - f_n^s)/f_n^p$ is $(2K_t/n\pi)^2$. For the fundamental mode in quartz, this difference is only about 0.37 per cent but with piezoelectric ceramics, which are characterized by much greater electromechanical coupling, the difference can be more like 10 per cent. An

x-cut quartz plate with a fundamental series frequency of 1 MHz needs to be about 2.9 mm thick while the corresponding thickness for a 1 MHz PZT-4 ceramic transducer is about 2.0 mm. With diameters of 50 mm, the clamped capacitances in these two examples are 27 pF and about 6 nF respectively. To avoid the use of very thin plates, frequencies above a few megahertz are usually obtained by driving the plate at a harmonic of the fundamental. In practice, it is necessary to hold the transducer in some way and, since the plane $x = 0$ is nodal when the device is operated at an odd-ordered resonance, it is best (when possible) to support the plate around the edge at $x = 0$.

Since we have ignored so far both internal damping in the piezoelectric material and the effects of the fluid around the transducer, the resonances have infinite Q at both extremes of load resistance. In practice, dielectric and frictional losses in the material provide some damping, even in vacuum, and limit the mechanical admittance to a finite value at resonance. For quartz these losses are very small, but piezoelectric ceramics often have sufficient internal damping to limit the Q to 10^3 or less. Also, with finite values of the load resistance, power is dissipated in the external circuit leading to an input mechanical resistance at resonance given by

$$R_\text{m}^\text{res} = (2C_0 h_{11}^2/\omega)\,[2\omega R_\text{L} C_0/(1 + \omega^2 R_\text{L}^2 C_0^2)] \qquad (5.4.25)$$

which reaches a maximum when $R_\text{L} = 1/\omega C_0$. This method of 'spoiling the Q' allows a much greater bandwidth to be obtained; with $R_\text{L} = 1/\omega C_0$, the frequency of maximum response is halfway between the series and parallel resonance frequencies, current damping limits the Q to $(n\pi/2K_t)^2$, and the bandwidth is approximately $f_n^p - f_n^s$. For the 1 MHz quartz transducer considered above, the load resistance corresponding to maximum current damping is 5.6 kΩ; this gives $Q = 270$ and a bandwidth of 3.7 kHz.

Reactive loads can be used to influence the mechanical properties of the plate. For example, a variable capacitance can be used to tune the transducer over the range from f^s to f^p. There is one special case worthy of mention; that is, when the load is a pure inductance L_0 chosen such that $L_0 = 1/(\omega^p)^2 C_0$, in which case $(1 + i\omega C_0 Z_e)$ vanishes and the input mechanical impedance goes to infinity at $\omega = \omega^p = 2\pi f^p$.

In practice, the rear face of the transducer will be in contact with a solid or fluid and the boundary condition there will actually be $\mathcal{T}_1(x = -l) = Z_a^b(\partial\xi/\partial t)$, where Z_a^b is the acoustical input impedance of the backing medium. While an exact solution for the coefficients in equation (5.4.17) is still possible, it is much more convenient to derive an approximate solution valid when $|(Z_a^b/\rho u)\cot(kl)| \ll 1$. With this approximation ξ_0 is given by

$$\xi_0 \sin(kl) = \frac{-(F/2\,\mathcal{Q}\omega\rho u)}{\cot(kl) - (K_t^2/kl)(1 + i\omega C_0 Z_e)^{-1} + i(Z_a^b/2\rho u)}. \qquad (5.4.26)$$

When the backing impedance is small compared with ρu, this expression will be a good approximation everywhere except close to the inactive resonances and even if the backing impedance is large, it will still hold close to the piezoelectrically active resonances of the plate. In fact we can again neglect ξ_1 entirely close to those resonances and approximate the input mechanical admittance by $-(i\omega/F)\xi_0 \sin(kl)$. At the frequency of the vacuum resonance, Y_m is therefore equal to $1/\alpha Z_a^b$ and this can be used to evaluate the reflection coefficient χ for normally incident plane waves arriving at the front face of the transducer from knowledge of the acoustical backing impedance.

It is now a simple matter to calculate the sensitivity of the transducer as a detector of sound arriving at one face. The electric flux density D_1, equal to $I/i\omega\alpha$, is given in terms of the relative displacement $\Delta\xi$ by equation (5.4.19) so that the current and voltage output of the transducer are given by

$$I = i\omega C_0 h_{11} \Delta\xi [1/(1 + i\omega C_0 Z_e)] \exp(i\omega t)$$
$$\xrightarrow[Z_e \to 0]{} i\omega C_0 h_{11} \Delta\xi \exp(i\omega t) \tag{5.4.27}$$

and

$$V = i\omega C_0 h_{11} \Delta\xi [Z_e/(1 + i\omega C_0 Z_e)] \exp(i\omega t)$$
$$\xrightarrow[Z_e \to \infty]{} h_{11} \Delta\xi \exp(i\omega t). \tag{5.4.28}$$

If the acoustic pressure at the front face of the transducer is p_a then the relative displacement $\Delta\xi = 2\xi_0 \sin(kl)$ may be obtained from equation (5.4.26) with $F/\alpha = p_a$; the open-circuit voltage sensitivity at resonance, V/p_a, then equals $2ih_{11}/\omega Z_a^b$. However, this mode of operation has some disadvantages. High sensitivity is achieved only with small values of the backing impedance but, under those conditions, the input mechanical impedance is not very large and the reflection coefficient may differ significantly from unity. However, if an inductance $L_0 = 1/\omega^2 C_0$ is placed in parallel with the transducer, so that the electrical network formed by L_0 and C_0 is resonant at the operating frequency, then the plate becomes a perfect pressure transducer with infinite input mechanical impedance but finite sensitivity.[9] From equations (5.4.26) and (5.4.28) the sensitivity under these conditions is given by

$$V/p_a = -(\alpha/C_0 h_{11}). \tag{5.4.29}$$

For example, a 1 MHz quartz transducer of diameter 50 mm operating with inductive tuning has $V/p_a = 17 \text{ mV Pa}^{-1}$ while, for the same transducer operating at low frequencies with rigid backing, $V/p_a = 2l(h_{11}/c_{11}^D) = 140 \text{ }\mu\text{V Pa}^{-1}$; these two values stand in the ratio $1 : K_t^2$. One can show that, when $Z_a^b \gg K_t^2 \rho u$, the sensitivity is at a maximum when the parallel inductance is equal to the value $1/\omega^2 C_0$; in practice, this provides an experimental criterion for tuning a variable inductor to resonance.

The transducer may also be operated as a source. If a harmonic potential difference $V \exp(i\omega t)$ is applied across the two faces then the crystal will vibrate mechanically in the thickness mode at the field frequency and radiate sound from both faces into the surrounding medium. To calculate the electrical input impedance of the transducer for these conditions, we again solve equations (5.4.11–13) but subject to different boundary conditions. If we let the acoustical impedance presented to the front face of the transducer be Z_a^f and that at the back continue to be Z_a^b then the boundary condition on the normal stress will be $\mathcal{T}_1 = -Z_a^f(\partial\xi/\partial t)$ at $x = l$ and $\mathcal{T}_1 = Z_a^b(\partial\xi/\partial t)$ at $x = -l$. The displacement will be given as before by equation (5.4.17) but, since the symmetric term $\xi_1 \cos(kx)$ does not couple to the electric field, we need consider only $\xi_0 \sin(kx)$. Again, an exact solution is possible but, as above, an approximation neglecting the terms $(Z_a^f/\rho u)\cot(kl)$ and $(Z_a^b/\rho u)\cot(kl)$ is much simpler:

$$\xi_0 \sin(kl) = \left(\frac{D_1(h_{11}/\omega\rho u^D)}{\cot(kl) + i[(Z_a^f + Z_a^b)/2\rho u^D]}\right). \qquad (5.4.30)$$

This expression is exact for any symmetric load ($Z_a^b = Z_a^f$), a good approximation for light asymmetric loading, and still holds near $kl = n(\pi/2)$ even under heavy loading. When equation (5.4.30) is combined with the expression for the potential difference across the transducer, (5.4.18), and the flux density is eliminated in favour of the current, the electrical input impedance is found to be

$$Z_e = \left(\frac{1}{i\omega C_0}\right)\left(1 - \frac{K_t^2/kl}{\cot(kl) + i[(Z_a^f + Z_a^b)/2\rho u^D]}\right) \qquad (5.4.31)$$

and the electrical admittance $Y_e = Z_e^{-1}$ to be

$$Y_e = i\omega C_0\left(1 + \frac{K_t^2/kl}{[\cot(kl) - (K_t^2/kl)] + i[(Z_a^f + Z_a^b)/2\rho u^D]}\right). \qquad (5.4.32)$$

$|Z_e|$ and $|Y_e|$ are plotted in figure 5.10 as a function of frequency for the 1 MHz quartz transducer considered above operating in unbounded argon at 0.1 MPa and 300 K. Since the diameter of the transducer is large compared with the wavelength of the radiated sound, the acoustical load impedances are purely resistive in this case and both equal to the product $\rho' u'$ of the fluid's density and sound speed; for gases this is very small compared with the characteristic impedance ρu^D of the transducer material. Under such light loading, $|Z_e|$ is minimum at a frequency very close to the series resonance frequency f_1^s and maximum very close to the parallel resonance frequency f_1^p. The frequency of maximum power output f_m is then equal to f_1^s when the transducer is driven by a voltage source (zero source resistance), or f_1^p when driven by a current source (infinite source resistance). For intermediate source resistances, f_m tracks the mechanical resonance frequency of the plate given by the solution of equation (5.4.22).

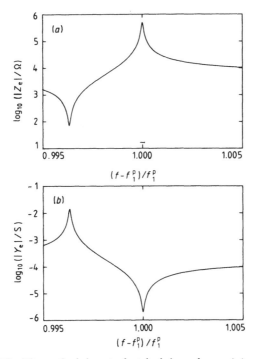

Figure 5.10 Theoretical input electrical impedance (a) and admittance (b) of a 1 MHz x-cut quartz transducer of diameter 50 mm operating in the thickness expander mode in unbounded argon at 300 K and 0.1 MPa.

Unless current damping is required to enhance the bandwidth, the usual approach is to drive the transducer from either a voltage or a current source at the corresponding resonance frequency. When constant-voltage excitation is used, the input electrical admittance is the most useful quantity and, for the transducer operating at the frequency of the unloaded series resonance $(\omega = \omega_n^s)$, this is given by the simple expression

$$Y_e = i\omega_n^s C_0 + \alpha_t^2 Y_m \tag{5.4.33}$$

in which Y_m is the inverse of the total mechanical load impedance presented to the transducer by the fluid: $Y_m = 1/\mathcal{A}(Z_a^f + Z_a^b)$. The transformation factor α_t^2 may be obtained using equations (5.4.16), (5.4.20) and (5.4.32):

$$\alpha_t = (\mathcal{A}h_{11}/l\beta_{11}^S). \tag{5.4.34}$$

It is sometimes convenient to place an inductance $L_0 = 1/(\omega_n^s)^2 C_0$ in parallel with the transducer to tune out the effect of the static capacitance C_0. This has no effect on the series resonance frequency because the electrical impedance seen by the transducer is the parallel combination of $i\omega L_0$ with the output resistance of the source (which is zero for an ideal voltage source). When tuned in this way, $Y_e = \alpha_t^2 Y_m$ and for the example case of the 1 MHz quartz transducer operating in unbounded argon (at 300 K and 0.1 MPa) $Y_e \approx 29$ mS ($Z_e \approx 35\ \Omega$) at ω_1^s.

In the case of constant-current operation, the electrical impedance is the most useful quantity and, at the frequency of the unloaded parallel resonance ($\omega = \omega_n^p$), this is given by

$$Z_e = (1/i\omega_n^p C_0) + (2h_{11}/\omega_n^p)^2 Y_m. \tag{5.4.35}$$

Note that the electrical input impedance is linear in the admittance and not the impedance of the mechanical load. The effect of the static capacitance can be tuned out with an inductor $L_0 = 1/(\omega_n^p)^2 C_0$ placed in series with the source; this also has no effect on the resonance frequency. For the example case considered above, Z_e is then equal to 1 MΩ at ω_1^p. Note that both equations (5.4.33) and (5.4.35) apply at the corresponding resonance frequency of the *unloaded* transducer. These are the relations between electrical and mechanical quantities exploited in the single-crystal interferometer.

For the source transducer of the double-crystal interferometer, constant surface displacement amplitude is usually sought despite variation in the acoustical load impedance. To achieve this, all one has to do is tune out the static capacitance with a parallel inductor and drive the combined network at a constant current I. The current flowing through the transducer is then $I/(1 - i\omega C_0 Z_e)$ so that, from equations (5.4.30) and (5.4.31), ξ_0 is constant and the displacement amplitude is given by

$$\xi_0 \sin(kl) = (I/i\omega C_0)(1/2h_{11}). \tag{5.4.36}$$

Alternatively, when the backing impedance Z_a^b is constant and large compared with the load impedance Z_a^f, constant-voltage excitation will do.

The quality factor for the transducer operating at the parallel resonance frequency in the unbounded fluid can be obtained from equation (5.4.32) by noting that $\cot(kl) \approx (n\pi/2) - kl$ near to resonance. This leads to a Lorentzian-type lineshape for $|Z_e|$ and gives

$$Q = (n\pi/4)(\rho u/\rho' u') \tag{5.4.37}$$

where ρ' and u' are the fluid's density and sound speed. For the 1 MHz x-cut quartz transducer operating in its fundamental mode against argon at 0.1 MPa and 300 K, $Q \approx 23\ 000$ and the bandwidth is about 43 Hz (neglecting other loss mechanisms). The same transducer operating at the $n = 3$ harmonic in water would have $Q \approx 24$ and a bandwidth of 125 kHz.

When operating near to a piezoelectrically active resonance, the electrical properties of the transducer can be modelled by the equivalence circuit shown in figure 5.11. The capacitance C_0 is, as above, the clamped capacitance of the transducer and is shown in parallel with a resistance R_0 representing the dielectric loss at the operating frequency (this loss is small for quartz but can be quite significant in piezoelectric ceramics). The inductance L_1, capacitance C_1 and resistance R_1 model the mechanical activity of the unloaded transducer and may be related to the lumped circuit effective mass, stiffness and resistance in vacuum. The impedance Z_L formed by the series combination of these circuit elements arises from loading imposed by the fluid and is equal to $\alpha(Z_a^f + Z_a^b)/\alpha_t^2$. In terms of the electrical parameters, the series and parallel resonance frequencies of the unloaded transducer are given by

$$f^s = (1/2\pi)(1/L_1C_1)^{1/2} \qquad (5.4.38)$$

and

$$f^p = (1/2\pi)[(C_0 + C_1)/(L_1C_0C_1)]^{1/2}. \qquad (5.4.39)$$

The equivalence circuit can easily be extended to cover operation of the transducer as a source by including in series with Z_L a voltage source generating $p_a l/(h_{11}/c_{11}^D)$ when sound impinges on one face. When both sides are exposed to sound, two voltage sources appear in series.

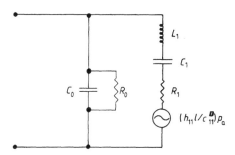

Figure 5.11 Equivalence circuit for a thickness-mode piezoelectric transducer.

In this theoretical treatment, a condition of one-dimensional strain has been assumed. However, with the finite diameter of real crystals, lateral clamping is not achieved in practice and other modes of the plate can be excited. This prevents the surface displacement from achieving the perfect piston-like motion of the model and leads to spurious resonances in the frequency spectrum. Practical methods of mounting the crystal also affect the performance. Nodal support is possible for reasonably large crystals and the Q can then approach the theoretical values given above. However, other methods of holding the crystal, such as edge clamping, introduce additional losses and may limit the vacuum Q to much less than theoretically possible. Nevertheless, with some care, excellent results can be achieved.

5.4.4 Piezoelectric flexure elements

The thickness expander, like other simple piezoelectric devices, has very high stiffness and consequently low static sensitivity and high resonance frequency. Much greater compliance, leading to high sensitivity and low resonance frequency, can be achieved by operating in a bending or flexure mode that exploits coupling between perpendicular fields.

For most purposes composite structures, such as those illustrated in figure 5.12, are required for flexure operation. The simplest composite is a bilaminar construction in which two piezoelectric strips are bonded together to form what is known as a bimorph. Figure 5.12(a) shows a bimorph cantilever fabricated from two piezoelectric ceramic strips poled in opposite directions. Application of a potential difference between the outer electrodes causes one strip to expand lengthways (3,1 coupling) while the other strip contracts. The differential strain developed in this way causes the cantilever to bend. The strength of the composite can be greatly enhanced by incorporating a metallic layer between the piezoelectric components as illustrated in figure 5.12(b). In this case the piezoelectric components can be poled in either the same or opposite directions; in the former case the central metal layer becomes one terminal of the device and the outer electrodes, connected in parallel, the other. The bonding techniques must be sufficiently good to prevent slippage of the layers during operation.

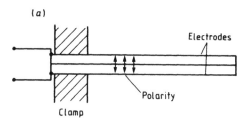

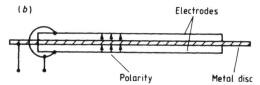

Figure 5.12 Piezoelectric flexure elements: (a) simple bimorph cantilever; (b) circular bimorph plate with central metal disc.

Calculation of the sensitivity and resonance frequency of a composite device is a rather complicated matter and it may be useful to have some approximate formulae based on practical results. The data in table 5.3 refer to bimorph strips fabricated from a commercial lead zirconate titanate piezoelectric ceramic (PXE5, Mullard plc) with a central metal layer of thickness $h/3$. Cantilever mounting and end-pinned support are both considered and the properties are given in terms of the free length l, the width w and the thickness h. When calculating the capacitance C for the cantilever the total length l' not the free length must be used. The quantity q/F gives the charge q developed under open-circuit conditions by application of a force F to the end of the cantilever or to the centre of the end-pinned strip. Conversely, the quantity z/q gives the displacement z generated (at the free end or at the centre) per unit applied charge. The approximate frequency f_0 of the fundamental resonance is also given and is valid when the width is small compared with the length.

Table 5.3 Properties of piezoelectric ceramic bimorphs [13].

Property	Cantilever support	End-pinned support
C/pF	$24\,[(wl'/h)/\mathrm{mm}]$	$24\,[(wl/h)/\mathrm{mm}]$
$(q/F)/(\mu\mathrm{C~N^{-1}})$	$0.35 \times 10^{-3}(l/h)^2$	$0.087 \times 10^{-3}(l/h)^2$
$(z/q)/(\mathrm{mm}\,\mu\mathrm{C^{-1}})$	$0.18\,[(l/wh)/\mathrm{mm^{-1}}]$	$0.045\,[(l/wh)/\mathrm{mm^{-1}}]$
f_0/kHz	$500\,[(h/l^2)/\mathrm{mm^{-1}}]$	$1400\,[(h/l^2)/\mathrm{mm^{-1}}]$

l = free length, w = width, h = thickness.

In transducer applications flexure elements are normally coupled to a diaphragm; two possible arrangements for doing this are illustrated in figure 5.13. The transducer in figure 5.13(a) is designed for non-resonant operation at audio frequencies and exploits a ceramic bimorph cantilever coupled to the centre of a diaphragm. This is the conventional design for 'ceramic' microphones. The edge-clamped circular metal plate shown in figure 5.13(b) has a piezoelectric ceramic disc bonded to its centre in a bilaminar construction that has useful properties both as a resonant transducer and in broad-band applications [13, 24]. The fundamental resonance frequency is typically quite close to that calculated for the unloaded plate. The device illustrated consists of a 10 mm × 1 mm PZT ceramic disc bonded to the centre of a 30 mm × 2 mm aluminium alloy diaphragm. The fundamental frequency of the transducer is about 18 kHz (compared with a calculated value of 21 kHz) and the first overtone is about 71 kHz. These frequencies are much lower than those easily achievable with thickness-mode transducers. Even when operating in a gas the quality factor is quite low (typically less than 50) and the transducer is suitable for pulsed as well as constant wave applications at its resonance frequency, and for wide-band non-resonant operation [14]. Bimorph discs of the type illustrated in figure 5.12(*b*) are also suitable for direct action on the fluid [21].

Piezoelectric polymer films can also be used as alternatives to ceramic in the fabrication of bilaminar flexure plate transducers. A simple tensioned diaphragm of the material, electroded on both surfaces, could also function as a microphone but, since the material would be stretched during both positive and negative deflections, the signal produced at the electrodes would be at twice the frequency of the incident sound wave. This square-law response could actually be advantageous in some applications. However, the usual approach is to form the polymer under slight tension over a curved backing surface made from a compliant material. Linear operation is obtained in this way because deflections of the film away from the backing increase the tension while motion in the opposite direction relaxes it. Piezoelectric polymer films are also available as bimorph laminates.

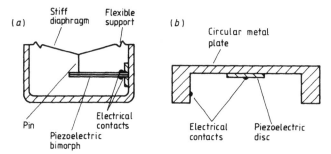

Figure 5.13 Transducers utilizing flexure elements: (*a*) audio frequency ceramic mircophone using bimorph cantilever; (*b*) edge-clamped metal plate with a plane piezoelectric disc bonded to its rear face.

The range of operating temperatures for composite flexure elements is not usually as wide as that for the parent piezoelectric material because of limitations in the bonding techniques. With ceramic materials, thin films of epoxy resin may be used below about 370 K but higher-temperature operation requires soldering of the major faces—an operation that can lead to thermal depoling of the ceramic if not carefully conducted. Quartz can of course be bonded at much higher temperatures but offers lower sensitivity. Since piezoelectric polymers are highly compliant, very stiff joints are not required and bonding is unlikely to limit the upper operating temperature significantly.

5.5 Capacitance Transducers

A typical capacitance transducer takes the form of a parallel-plate capacitor in which one plate, the back plate, is massive while the other is a light diaphragm that moves in response to a mechanical or electrostatic force. The diaphragm may be a tensioned metal-foil membrane, a tensioned metallized polymer film or an edge-clamped metal plate. To operate the transducer as a source of sound, an AC signal (often combined with a DC bias voltage) is applied across the plates and generates an electrostatic force between them that excites motion of the diaphragm. When operated as a detector, the incident sound waves deflect the diaphragm thereby modulating the capacitance and, under conditions of constant stored charge, generating an alternating potential difference between the plates. Capacitance transducers are wide-band devices usually operated at frequencies below that of the fundamental resonance of the diaphragm. Although it is possible to design sealed 'submersible' devices for use in liquids, capacitance transducers are best suited to operation in gases.

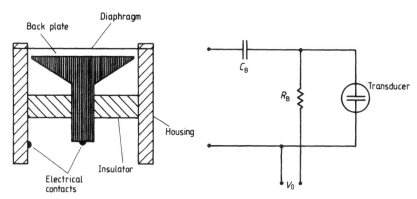

Figure 5.14 Schematic construction of capacitance transducer (left) and a basic circuit of applying DC bias (right). The conducting surface of the diaphragm is in electrical contact with the metal housing.

A typical arrangement is shown in figure 5.14 together with a basic circuit used to impose a bias voltage V_0 across the transducer. The blocking capacitor C_B prevents the bias voltage from appearing across the signal terminal of the transducer, and the blocking resistor R_B prevents alternating current from flowing in the bias source. Since R_B and C_B are chosen such that $R_B \gg 1/\omega C$ and $C_B \gg C$, where C is the active capacitance of the transducer, the bias circuit has negligible influence on the performance of the network at signal frequencies. Assuming that the membrane consists of a metallized dielectric film of thickness d_1 and dielectric constant ε_r separated from the back plate by a gap d_0, the electric field in the gap is

$$E_0 = V_0/(d_0 + d_1/\varepsilon_r). \qquad (5.5.1)$$

The case of an all-metal membrane corresponds to $d_1 = 0$ so that the field reduces to V_0/d_0.

5.5.1 Capacitance microphones

The open-circuit sensitivity of the transducer as a detector is given by

$$V/p_a = E_0(\langle \xi \rangle / p_a) \qquad (5.5.2)$$

where $\langle \xi \rangle$ is the mean surface displacement amplitude. Constant charge operation is achieved in practice by connection to an amplifier that has a very high input resistance. Since the capacitance of the transducer is usually small (< 100 pF), even a small capacitance C_1 loading the output can be a serious problem through division of the output signal by the ratio $(C + C_1)/C$. To overcome this problem a preamplifier is normally located directly behind the back plate so that cable capacitance is eliminated. The output impedance of the preamplifier is quite low (≈ 1 kΩ or less) so that quite long coaxial cables may be driven. The simple circuit of figure 5.15(a) using a single FET is quite adequate for most purposes. The noise level in the circuit is controlled mainly by the gate resistance and it is worthwhile striving for the highest possible value here. Since DC offset is unimportant and JFET gate leakage currents can be very small anyway, gate resistors of order 1 GΩ or more can be used. A slightly more complicated alternative using a remote unity-gain preamplifier and active guard is illustrated in figure 5.15(b). This scheme has the advantage that the maximum operating temperature of the transducer is not limited by the electronics. A 'triaxial' cable is used in which the centre conductor, which carries the signal from the transducer, is screened from the outer shield (ground) by a coaxial guard on to which is fed the output from the preamplifier. Since this preamplifier has a voltage gain G close to unity, there is no potential difference across the capacitance C_1 between the inner conductor and ground, and the effective value at signal frequencies is reduced to $(1 - G)C_1$. The circuit shown in figure 5.15(b) also uses this bootstrap technique (via C_D) to reduce the gate–drain capacitance [19]. Actual preamplifiers constructed with high input resistance FETs achieve gains between 0.99 and 0.999 so that capacitances of order 100 pF can be tolerated.

Capacitance transducers are very sensitive devices. The intrinsic noise level is extremely small and in practice the noise in the system is usually controlled by the electronics. For the JFET circuit illustrated in figure 5.15(a), the noise voltage density at signal frequencies is given approximately by

$$e_n = [e_a^2 + (4kT/R_{in}\omega^2 C^2)]^{1/2} \qquad (5.5.3)$$

where e_a is the noise voltage density of the JFET (and any subsequent amplifier stages) referred to the input and the second term in the equation is the Johnson noise developed in the gate resistance R_{in}.[10] At low frequencies this latter noise source dominates over the intrinsic noise of the amplifier. For example, with $C = 10$ pF and $R_{in} = 1$ GΩ, Johnson noise in the gate resistor leads to $e_n \approx 60$ nV Hz$^{-1/2}$ at 1 kHz. As an example, consider a typical capacitance transducer for audio frequency applications constructed with a nickel-foil diaphragm 5 μm thick and 9 mm in diameter stretched to a tension of 3200 N m^{-1} [5, 15]; this gives a fundamental resonance frequency near 23 kHz. The back plate is separated from the diaphragm by a static gap of about 20 μm and the microphone is operated with a 200 V DC bias voltage. The low-frequency displacement sensitivity, averaged over the surface of the membrane, is (from equation (5.3.5)) about 0.8 nm Pa^{-1} so that the open-circuit voltage sensitivity is about 8 mV Pa^{-1}. With a noise floor in the electronics of, say, 0.1 μV RMS (easily achievable with lock-in detection), sound pressure levels down to around 10 μPa are detectable at which level the average displacement amplitude of the diaphragm is of order 10^{-14} m!

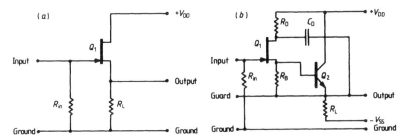

Figure 5.15 Preamplifier circuits utilizing *n*-channel JFETs for use with capacitance microphones: (*a*) an elementary circuit with voltage gain $R_L g_m/(1 + R_L g_m)$, input resistance R_{in}, and output resistance $1/g_m$ (g_m is the transconductance of Q_1); (*b*) Unity-gain circuit with active load and bootstraps to raise input impedance [19].

A few other details of transducer design may be useful. To prevent excessive enhancement of the membrane stiffness by the gas trapped in the space between the diaphragm and the back plate, holes are usually provided in the back plate to allow access to a relatively large buffer volume (back volume). Viscous flow resistance to and through these holes constitutes the main damping mechanism in the microphone and usually results in a low Q for the fundamental resonance. Some mechanism for pressure equalization between the inside of the transducer and the external environment must also be provided. However, the 'leak' through which this takes place will then determine the lower-frequency limit of the device as an acoustic pressure transducer.

5.5.2 Reversible operation

Capacitance transducers also make good broad-band generators of sound for acoustic experiments. When an alternating signal $V_1 \cos(\omega t)$ is applied the total electrostatic force between the plates is

$$F = - \left(\frac{\varepsilon_0 \mathcal{Q}}{2}\right) \left(\frac{V_0 + V_1 \cos(\omega t)}{d_0 + d_1/\varepsilon_r}\right)^2 . \tag{5.5.4}$$

Because this force depends upon the square of the potential difference, sound can be produced both at the applied angular frequency ω and also at 2ω. Usually, the bias voltage V_0 is large and $V_1 \ll V_0$ so that the drive frequency $\omega/2\pi$ predominates, but if $V_0 = 0$ then sound is produced only at twice the applied frequency. This frequency doubling can be used to advantage where direct electromagnetic coupling ('crosstalk') between source and detector is a problem; with the source driven at half the desired acoustic frequency, crosstalk can be filtered electronically from the received signal.

5.5.3 Solid dielectric capacitance transducers

The type of capacitance transducer illustrated in figure 5.14, with a stiff diaphragm separated from the back plate by a gas-filled gap, is by no means the only possible kind. An important variant, from the perspective of the kind of measurements with which this book is concerned, is the so-called solid dielectric capacitance transducer [22]. This employs a thin dielectric membrane (typically a polymer about 10 μm thick) which lies flat over the surface of the back plate under minimal tension. The outer surface of the membrane is metallized to form the second conducting plate of the capacitor. The mode of operation is still the same but now the stiffness of the system is provided almost entirely by the gas trapped in the small gap between the plates. The effective mass is determined by a combination of the membrane mass and the mass of the fluid near to the moving surface. When the plate is smooth, the trapped volume is very small and the stiffness is correspondingly large. This results in a wide bandwidth over which the sensitivity of the device is uniform (though small). To enhance the sensitivity, the stiffness can be reduced by roughening the back plate, by machining grooves or blind holes in it, or by providing vent holes to a back volume.

The displacement amplitude $\langle \xi \rangle$ of the membrane under alternating pressure p (which may be acoustic or electrostatic in origin) may be calculated from an estimate of the mechanical impedance using equation (5.3.11). Denoting the volume available to the gas behind the membrane by v_0, the stiffness may be written

$$s = (\pi a^2)^2 (\gamma_{\text{eff}} / \varkappa_T v_0) \tag{5.5.5}$$

where $\varkappa_T / \gamma_{\text{eff}}$ is an effective compressibility for the enclosed gas.[11] Since the gap between the plates is very small, the gas there will remain very nearly isothermal during expansion and compression resulting from the motion of the membrane. Consequently, γ_{eff} approaches unity unless there are significant buffer volumes with dimensions larger than the thermal penetration length δ_{h}. The corresponding low-frequency sensitivity of the transducer as a microphone with DC bias V_0 is

$$V / p_{\text{a}} = [V_0 \varepsilon_{\text{r}} / (d_1 + \varepsilon_{\text{r}} d_0)] (v_0 \varkappa_T / \gamma_{\text{eff}} \pi a^2) \tag{5.5.6}$$

where d_0 is now just the effective gap arising from the surface roughness of the back plate. Neglecting the mass of the fluid, the effective mass is equal to the true mass and the resonance frequency is

$$f_{\text{res}} = (1/2\pi)(\pi a^2 \gamma_{\text{eff}} / \varkappa_T \rho_S v_0)^{1/2} \tag{5.5.7}$$

which simplifies to $(1/2\pi)(p/\rho_S d_0)^{1/2}$ when the gas is perfect and there are no additional buffer volumes. For a device of 10 mm diameter with a 12 μm thick polymer-film membrane ($\rho_S \approx 15$ mg m^{-2}, $\varepsilon_{\text{r}} \approx 2$) and a back volume of 10 mm^3 operating at $p = 0.1$ MPa the resonance frequency would be about 40 kHz; the low-frequency sensitivity with 200 V DC bias would be about 10 mV Pa^{-1}. If the back volume were eliminated and the residual volume in the surface roughness were equivalent to a mean gap of 1 μm then the resonance frequency would increase to around 400 kHz and the sensitivity decline to about 0.1 mV Pa^{-1}.

Solid dielectric capacitance transducers are easy to fabricate, rugged, and of high sensitivity. At moderate temperatures, metallized polymer films make the best membranes [23]. A few alternatives remain for high-temperature operation; for example, plated glass or mica sheets can be used [16].

5.5.4 Electret materials

Electrets are materials which have 'permanent' electric charge. They include polarized materials (dipolar electrets) and materials containing trapped charges of either or both sign. When fabricated in the form of thin films with the direction of polarization perpendicular to the plane of the material, they become useful materials for the construction of capacitance transducers. The advantage of using an electret material as the diaphragm is that no DC bias voltage is required; the polarized film generates its own electric field. Electret membranes can be used either as tensioned diaphragms separated from the back plate by a gap or as untensioned membranes in solid dielectric capacitance transducers. Alternatively, the electret material may be inserted as a separate layer between the back plate and diaphragm.

Commonly, electret materials for transducer applications are polymer films such as polyvinylidene fluoride (PVDF), polyethylene terephthalate (PET, Mylar), polytetrafluoroethylene (PTFE, Teflon-TFE), and the copolymers tetrafluoroethylene-hexafluoropropylene (Teflon-FEP) and tetrafluoroethylene-perfluoromethoxyethylene (Teflon-PFA). There are several mechanisms by which the electret effect may be introduced. Dipolar electrets (e.g. PVDF) are polarized by alignment of elementary dipoles in an electric field at elevated temperature. Internal space charges can also be polarized in this way, or space charges can be injected by application of high fields ($\sim 10^7$ V m^{-1}). It is also possible to make monopolar materials by depositing charge in the material near to one surface using corona discharge in air or an electron beam in vacuum; this is actually the method used to produce most commercial electrets.

In the usual model of monopolar electret materials, one assumes that an effective charge density σ is held in a layer just below the back surface of the electret. The electric field in the gap between the film and the back plate is then given by

$$E_0 = (\sigma d_1/\varepsilon_0)(d_1 + \varepsilon_r d_0)^{-1} \qquad (5.5.8)$$

where ε_0 is the electric permittivity of the gas in the gap. Comparing this with equation (5.5.1) we see that an electret charge density of around 160 μC m^{-2} is equivalent, for a typical transducer ($d_0 = 20$ μm, $d_1 = 12$ μm, $\varepsilon_r = 2.2$), to a DC bias of about 100 V. Actual charge distributions in electret films can differ substantially from this simple model of surface charge retention so that σ should be viewed very much as an 'effective' charge density. In practice values in the range 100 to 200 μC m^{-2} are achieved [6].

Electrets are remarkably stable materials in most respects but they do suffer from thermal depoling at elevated temperatures. PVDF is claimed to depole rapidly above 370 K while the corresponding temperature for PTFE is around 500 K [17]. These figures should be taken only as a guide; actual performance depends to some extent on the method of polarization and the chemical environment in which the electret is used. For example, contact with common organic liquids can cause rapid loss of polarization at ambient temperatures. The most stable electret polymers in current use are Teflon and the Teflon copolymers which achieve optimum charge retention after annealing at temperatures of around 470 K and repeated corona charging. These materials can have long charge retention times at temperatures up to at least 420 K [17]. Occasionally the electret effect causes problems in nominally unpolarized dielectrics used in transducers with DC bias. Mylar is frequently used in such transducers but, operating under DC bias, it tends to acquire polarization even at ambient temperatures. Since the induced polarization is in opposition to the applied field, the efficiency of the transducer is diminished. One solution is simply to turn off the applied field and operate the transducer as an electret although it is possible to end up with a sample that discharges too fast to be useful as an electret but which reacquires polarization in an electric field too rapidly to be effective under DC bias.

5.6 Moving-coil Transducers

The moving-coil principle is widely exploited in the manufacture of general-purpose microphones and loudspeakers for audio frequency applications. The basic arrangement, illustrated in figure 5.16, consists of a stiff diaphragm which supports a circular coil within the radial field of a permanent magnet. Motion of the diaphragm causes the coil to move perpendicular to the field, thus generating an open-circuit voltage V given by

$$V = Blv. \tag{5.6.1}$$

Here, B is the magnetic flux density, l is the length of wire in the coil, and v is the speed. In order to obtain a uniform broad-band response as a detector, the speed amplitude must be independent of frequency; this requires its motion to be controlled by a mechanical or acoustical resistance. Resistance control is usually achieved by some method of restricting the flow of fluid to and from the space behind the diaphragm so that viscous losses damp the motion. The device illustrated in figure 5.16 employs a porous material (e.g. silk cloth) for this purpose. Another method that has been used involves fabricating the coil former from an electrically conducting material so that its motion in the magnetic field is damped by eddy currents. Although damping can be successful in flattening the frequency response of the transducer, the open-circuit sensitivity as a detector is quite small (around $1\,\text{mV}\,\text{Pa}^{-1}$ is typical). However, the source impedance is also small (typically a few hundred ohms) and the signal-to-noise ratio compares favourably with other types of transducer. To keep both the mass and the electrical resistance of the coil small, it is usually wound from aluminium wire or ribbon, insulated by anodizing or enamelling, rather than from copper (which has a greater product of mass and electrical resistance). Typical diaphragm materials are aluminium alloys, plastics or paper. By forming the centre of the diaphragm into a dome, for rigidity, and the edge into concentric corrugations, for flexibility, piston-like motion of the central portion can be achieved over a substantial frequency range. Composite structures can also be used to obtain a stiff central region but a compliant support.

Moving-coil transducer may also be designed for operation at a fixed frequency. Typically the diaphragm is then a flat edge-clamped plate operated at its fundamental resonance frequency without deliberate enhancement of the mechanical resistance. The greater stiffness of this arrangement allows higher resonance frequencies than can be achieved with the corrugated edge.

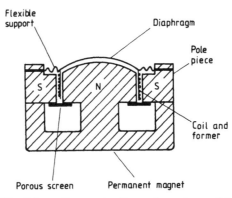

Figure 5.16   Schematic construction of moving-coil transducer.

Either kind of transducer may also be operated as a source of sound. A current I flowing in the coil produces a force of magnitude BlI; consequently, at frequencies such that the diaphragm moves as a single unit, the speed amplitude is given by

$$v = BlI/Z_m \qquad (5.6.2)$$

where Z_m is the mechanical impedance. Because of the large damping needed to achieve a flat frequency response, the power dissipation in such a moving-coil source can be quite large.

Many of the complicated factors important in the design of good microphones and loudspeakers are irrelevant to experimental applications of moving-coil devices in resonators and interferometers. Instead, one must face the problems of choosing materials and damping techniques that will function under the experimental conditions. In particular, adhesives are usually required to bond parts of the coil assembly and these will limit high-temperature applications; also, while it is possible to optimize viscous losses for uniform frequency response at one temperature and pressure, the damping will change significantly with the nature and state of the surrounding fluid.

5.7 Photoacoustic Excitation

The electroacoustic transducers discussed above work by direct mechanical action on the fluid. Photoacoustic excitation offers an alternative method of generating sound waves in fluids without any such direct mechanical coupling being required. The sample is irradiated with a beam of light, the frequency of which is chosen to match one of the absorption bands of the molecules in the fluid. Thus, the energy absorbed by the system feeds directly an internal mode of the molecules but, through collisions with other molecules, it can be shared rapidly with the translational modes causing local heating and a commensurate increase in pressure. In order to generate sound by this method, all one has to do is modulate the radiation at the required frequency. Typically, the radiation used lies either in the infrared region (for excitation through a vibrational absorption band) or in the visible or ultraviolet regions (for excitation through an electronic absorption band). The radiation need not be highly monochromatic, although lasers (which are) may form a very convenient source. For example, the common helium–neon laser emits radiation with a wavelength close to the 3.39 μm absorption band of methane and has been used for excitation of acoustic resonators filled with that gas [18]. In other cases, tunable lasers or incandescent sources may be required. Modulation may be achieved using a mechanical chopper or by more sophisticated means such as pulsed operation of the source.

Two limiting cases may be discerned. In the first, the optical absorption coefficient in the fluid is sufficiently small that the intensity of the beam does not diminish significantly across the optical path. The modulated beam then acts as a line source and may be used to excite selectively the normal modes of an acoustic cavity. For example, such a beam directed along the axis of a cylindrical resonator will be efficient in exciting radial modes but not in exciting azimuthal modes (for which the on-axis acoustic pressure vanishes). This kind of selectivity is probably the main advantage offered by photoacoustic excitation. Even though only a small fraction of the incident power may be absorbed, the method is effective because typical power levels required in acoustic experiments (< 1 μW) are small compared with optical power levels (which can easily exceed 1 mW). In the second case, the optical absorption coefficient is high and the beam, being attenuated rapidly in the fluid, acts more like a point source of sound and excites many modes simultaneously.

5.8 Waveguides

The requirement for transducers to form an integral part of an acoustic resonator or interferometer can be a demanding one, especially if the experiment is to be operated at high temperatures. However, an alternative approach is possible in which remote transducers, operated at a convenient temperature, are coupled to the resonator through waveguides consisting of small cross-section hollow tubes.[12] The remote end of the tube may terminate in a small chamber containing the transducer or at the more or less rigid surface of the transducer itself. As well as separating transducers from the possibly harsh conditions to which the resonator itself may be exposed, waveguides allow much greater flexibility in size and shape of the transducers and may, for example, allow commercially available devices to be employed in place of ones fabricated specially for the purpose. Another advantage is that source and detector transducers can be physically separated by much greater distances than otherwise possible so that direct electromagnetic or even mechanical coupling between them may be diminished. But, weighing against these advantages, one must address the difficulties in obtaining acceptable signal levels while simultaneously keeping perturbations to the behaviour of the resonator small. Actual absorption losses along the waveguide need not be very large but often the coupling efficiency is poor because of improperly matched impedances and can only be improved at the expense of increased perturbations to the performance of the resonator. Mainly for this reason, waveguides have not yet been widely used. However, it seems likely that, if high-precision acoustic methods are to be used in the future at elevated temperatures, this kind of technology will have to be exploited. Accordingly, this section is directed towards mathematical models suitable for the design and operation of waveguides. The questions to be addressed are twofold: first, what is the perturbation to the resonance frequencies and linewidths of the acoustic resonator caused by the presence of the waveguide; and, second, will an acceptable signal level be obtained?

5.8.1 Admittance of opening

In order to calculate the effect of any opening in the wall of an acoustic resonator, the specific acoustic admittance $\rho u(v/p_a)$ there is required. Narrow tubes are generally employed as waveguides and operated below the first cut-off frequency where only plane waves can propagate. In this case, a theoretical treatment of the problem is relatively simple: the specific acoustic admittance is constant over any cross-section of the tube and its value at the opening, y_0, may be expressed in terms of the value y_L at the termination of the tube and the propagation constant for plane waves travelling along the wave guide. The standing wave pattern within the tube can then be obtained by finding the superposition of positive- and negative-going simple-harmonic plane waves that satisfies the boundary conditions at the ends.

Let the coordinate z measure the distance from the opening in the wall of the resonator; the acoustic pressure for the positive-going wave will then be $p_+ \exp(-ik_{KH}z)$ while for the negative-going wave it will be $p_- \exp(ik_{KH}z)$. Here, we have allowed for losses at the wall of the tube by setting the propagation constant equal to the Kirchhoff–Helmholtz value $k_{KH} = (\omega/u) + (1 - i)\alpha_{KH}$. The total acoustic pressure and fluid velocity are therefore given by

$$p_a = A \cosh(\Phi_0 - ik_{KH}z) \tag{5.8.1}$$

$$v = (A/\rho u)\sinh(\Phi_0 - ik_{KH}z) \tag{5.8.2}$$

where $A_+ = \frac{1}{2}A \exp(\Phi_0)$ and $A_- = \frac{1}{2}A \exp(-\Phi_0)$, and the dimensionless ratio $y = \rho u(v/p_a)$ is

$$y = \tanh(\Phi_0 - ik_{KH}z). \tag{5.8.3}$$

We now write the admittance at $z = L$ as

$$y_L = \tanh(\Phi_L) = -i \tan(i\Phi_L). \tag{5.8.4}$$

Since the complex reflection coefficient χ_{Re} is equal to $(1 - y_L)/(1 + y_L)$ in this notation, we see that $\chi_{Re} = \exp(-2\Phi_L)$.[13] Comparing equations (5.8.3) and (5.8.4), we see that the parameters Φ_0 and Φ_L are related by

$$\Phi_0 = \Phi_L + ik_{KH}L \tag{5.8.5}$$

and that y_0 is given by

$$y_0 = \tanh(\Phi_L + ik_{KH}L) = i \tan(k_{KH}L - i\Phi_L)$$
$$= i[\tan(k_{KH}L) - \tan(i\Phi_L)]/[1 + \tan(k_{KH}L)\tan(i\Phi_L)]. \tag{5.8.6}$$

This equation gives y_0 in terms of the parameter Φ_L that characterizes the terminal admittance.

5.8.2 The closed tube

If the tube is closed at $z = L$ by a rigid termination at which the reflection coefficient is unity ($y_L = 0$) then the specific acoustic admittance is zero there but equal to

$$y_0 = \tanh(ik_{KH}L) = i \tan(k_{KH}L) \tag{5.8.7}$$

at the opening. Provided that the dissipation is small ($\alpha_{KH}L \ll 1$), $|y_0|$ exhibits sharp minima and maxima: $\text{Re}(k_{KH}L) = n\pi$ at the maxima while $\text{Re}(k_{KH}L) = (n + \frac{1}{2})\pi$ at the minima ($n = 0, 1, 2 \cdots$). Increasing $\alpha_{KH}L$ causes the ratio of the maximum and minimum values of $|y_0|$ to diminish and ultimately, for large values of $\alpha_{KH}L$, y_0 approaches unity and all the acoustic energy entering the waveguide at $z = 0$ is absorbed. Thus, for efficient operation of the waveguide, $\alpha_{KH}L$ must be small. When y_L is small but not zero, the perturbation caused to the resonance frequencies of the cavity by the presence of the waveguide will still be minimized when $\text{Re}(k_{KH}L)$ is near $n\pi$ with integer n. When the length of the tube satisfies that condition, the admittance of the opening will be

$$y_0 = (y_L + \alpha_{KH}L)/(1 + y_L\alpha_{KH}L) \approx y_L + \alpha_{KH}L \qquad (5.8.8)$$

so that, tuned in this way, a lossless waveguide projects an acoustic image of the termination on to the opening: $y_0 \approx y_L$.

Figure 5.17 shows the behaviour of y_0 as a function of $\text{Re}(k_{KH}L/\pi)$ for a rigidly closed tube of fixed length 20 mm and diameter 1 mm operating in argon at 300 K and 0.1 MPa. The speed of sound under these conditions is about $320\ \text{m s}^{-1}$ so that the frequency range illustrated extends up to 20 kHz. In this range, $\alpha_{KH}L$ is small but not negligible.[14]

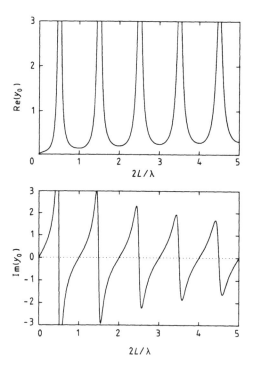

Figure 5.17 Input specific acoustic admittance y_0 of a cylindrical waveguide of length $L = 20$ mm and diameter 1 mm, rigidly terminated at $z = L$, as a function of $\text{Re}(k_{KH}L/\pi) = 2L/\lambda$. The medium is argon at 300 K and 0.1 MPa.

5.8.3 *Tunable waveguides and waveguides of variable cross-section*

When the resonator is to be operated over a range of frequencies, the perturbation caused by the presence of a waveguide will vary strongly with the frequency and may be unacceptably large for values of $\text{Re}(k_{\text{KH}}L)$ near to $(n + \frac{1}{2})\pi$. One solution to this problem is the use of a tunable waveguide, the length of which is adjusted so that y_0 is always minimized. If we include dissipation in the waveguide but assume that $y_L = 0$ then the acoustic pressure at the face of the transducer will be given by

$$p_L = p_0/\cos(k_{\text{KH}}L) \qquad (5.8.9)$$

where p_0 is the acoustic pressure at the open end of the tube; for the moment we assume that the transducer is not being used to generate sound but only to detect a sound field that exists outside the waveguide.

In order to calculate p_0, it is necessary to study the boundary conditions at the open end of the tube in a little more detail. The reason for this is that the presence of the opening in the wall of the acoustic cavity will perturb the sound field there so that p_0 will differ from the acoustic pressure that would exist in the absence of the opening. Let us first consider a simplified model in which the waveguide opens into the infinite half space $z < 0$ through a rigid baffle, and let the sound field in this half space consist of normally incident plane waves arriving from $z = -\infty$, plus whatever waves are reflected back from the surface $z = 0$. If the incident wave is $p_i \exp(-ikL)$ then, in the absence of the opening in the baffle, the reflected wave would be $p_i \exp(ikL)$ and the acoustic pressure everywhere on the surface $2p_i$. In the presence of the waveguide, the specific acoustic admittance generally differs from zero over the area of the opening so that the normal fluid speed v does not vanish there. However, in our present approximation, v is at least constant over the area of the opening so that in cylindrical polar coordinates (z, r, θ) coaxial with the tube we have

$$
\begin{aligned}
v &= v_0 & r \leqslant b, z = 0 \\
v &= 0 & r > b, z = 0.
\end{aligned}
\qquad (5.8.10)
$$

The acoustic pressure in the half space $z < 0$ is then given by

$$p_a(z, r, \theta) = p_i[\exp(ikz) + \exp(-ikz)] + p_{\text{rad}}(z, r, \theta) \qquad (5.8.11)$$

where $p_{\text{rad}}(z, r, \theta)$ is an additional wave radiated from the open end of the waveguide and chosen to satisfy the boundary conditions. Since only plane waves may propagate in the tube, this radiated wave will be of the same form as one emitted from a vibrating circular piston set in the baffle. Notice that the form of equation (5.8.11) automatically ensures that the acoustic pressure is a continuous function of position and that, since the normal derivative of the term in brackets vanishes for $z = 0$, the fluid flow into and out of the open end of the tube is associated only with the radiation term. The boundary conditions at the open end of the waveguide require that the total acoustic pressure satisfy

$$p_0 = \rho u v_0 z_0 \qquad (5.8.12)$$

and that the radiation term component of the acoustic pressure, $p_0 - 2p_i$, satisfy

$$p_0 - 2p_i = \rho u v_0 z_r. \qquad (5.8.13)$$

Here, z_0 is the specific acoustic impedance of the opening and z_r is the specific acoustic radiation impedance of a circular piston of radius b. Eliminating v_0 between equations (5.8.12) and (5.8.13), we find

$$p_0 = 2p_i [z_0/(z_0 - z_r)]$$
$$= 2p_i/[1 - iz_r \tan(k_{KH}L)] \qquad (5.8.14)$$

and hence that the acoustic pressure at the face of the transducer is

$$p_L = 2p_i/[\cos(k_{KH}L) - iz_r \sin(k_{KH}L)]. \qquad (5.8.15)$$

If we neglect the radiation impedance altogether we assume, in effect, that the wave field is unaffected by the presence of the opening so that $p_0 = 2p_i$. This is a fair approximation when $kb \ll 1$. With that approximation and neglecting losses along the waveguide, the acoustic pressure at the face of the transducer varies periodically like $\sec(kL)$ and the condition for $y_0 = 0$ corresponds to the values of kL at which $|p_L|$ is also a minimum: $kL = n\pi$, $n = 0, 1, 2 \cdots$; and $p_L = (-1)^n 2p_i$. For intermediate values, $kL = (n + \frac{1}{2})\pi$, the tube is resonant and both y_0 and p_L diverge to $\pm\infty$. It is therefore possible to tune the waveguide for minimum perturbation by locating a length corresponding to minimum received signal amplitude. When tuned in this way, an acoustic image of the transducer is placed flush with the plane $z = 0$ so that the received signal is the same as one could measure with a truly flush detector.

Two factors will cause the actual performance of a tunable waveguide system to fall short of this ideal. First, a practical waveguide is likely to be quite long so that $\alpha_{KH}L$ is not entirely negligible. Second, the radiation term perturbs the sound field near to the opening of the waveguide resulting in a shift between the lengths L at which p_L is minimum and those at which y_0 is minimum. Provided that kb is small, z_r may be approximated using equations (5.2.29) which, when combined with (5.8.15), show that the values of $\mathrm{Re}(k_{KH}L)$ corresponding to the maxima and minima of $\mathrm{Re}(p_L)$ are all increased by $(8/3\pi)(kb)$ in leading order, while those corresponding to maxima and minima of $\mathrm{Im}(y_0)$ remain unchanged. With z_r non-zero, we also see from equation (5.8.14) that the acoustic pressure at the opening of the waveguide will differ from $2p_i$ when the frequency is near to one of the resonance frequencies of the tube. However, at frequencies remote from resonance, radiation from the open end of the tube affects the sound field there but little when $|z_r| \ll 1$.

As an illustration, figure 5.18 shows the magnitude of the acoustic pressure at both $z = L$ (full curve) and $z = 0$ (broken curve) as a function of length for a 3 mm diameter tunable waveguide operating in argon at 300 K and 0.1 MPa with a fixed frequency of 5 kHz ($k \approx 100 \text{ m}^{-1}$, $\alpha_{KH} = 1.8 \text{ m}^{-1}$). When tuned near $\text{Re}(k_{KH}L) = 10\pi$, the maximum length in this example (about 0.32 m), the dissipation is sufficient to reduce the acoustic pressure at the termination to 85 per cent of its value at the opening. The behaviour of p_0 shows that radiation is quite important when the ratio L/b is relatively small (< 10) and the waveguide is imperfectly tuned but, for lengths great compared with the radius, the fluctuations of p_0 are damped out and radiation is of secondary importance to dissipation. Figure 5.19 shows the behaviour of y_0 for the same set of conditions. Here, the locus of minima of the specific acoustic conductance is given by $\text{Re}(y_0) = \alpha_{KH}L$ and reaches 0.57 at the maximum length illustrated. Thus, although the imaginary part of y_0 can be tuned out quite efficiently, there will still be a loss of energy from the external sound field which may be significant. Notice that, if values of $|y_0|$ of order unity lead to acceptably small perturbations, tuning eventually becomes unnecessary when the waveguide is sufficiently long. As figures 5.18 and 5.19 show, this condition can be achieved without great loss of signal.

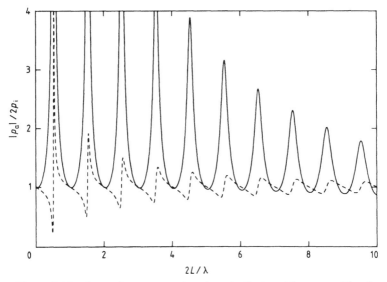

Figure 5.18 Acoustic pressure at each end of a tunable waveguide of diameter 3 mm (operating in argon at 300 K and 0.1 MPa) as a function of $\text{Re}(k_{KH}L/\pi) = 2L/\lambda$. The frequency is constant at 5 kHz and the length varies up to 0.32 m. ————, $|p_L|$; – – – –, $|p_0|$.

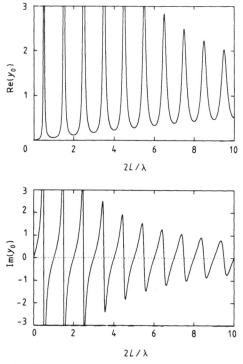

Figure 5.19 Specific acoustic admittance of the tunable waveguide with conditions as in figure 5.18.

Most of the conclusions reached above apply equally well to a waveguide opening into a resonator (rather than into an infinite half space) provided that we interpret $2p_i$ as simply the unperturbed acoustic pressure at the opening. When the acoustic resonator is operated near to one of its resonance frequencies, the radiation impedance will be modified by the addition of a resonance term which may be of substantial magnitude compared with the reactance and equation (5.2.37) applies.

Of course there is one condition that we have not yet considered, namely the case in which the external sound field is driven by a transducer at $z = L$ coupled through the waveguide. In this case the boundary conditions require that the fluid speed in the tube at $z = L$ equal that of the vibrating surface of the transducer and that the ratio $p_0/\rho u v_0$ equal the radiation impedance of a circular piston of radius b; these are sufficient to determine the two constants A and Φ_0 of equations (5.8.1) and (5.8.2). The acoustic pressure and fluid speed in the waveguide are thus given by

$$p_a = \rho u v_L \left[\cosh(\Phi_0 - ik_{KH}z)/\sinh(\Phi_0 - ik_{KH}L)\right]$$
$$v = v_L \left[\sinh(\Phi_0 - ik_{KH}z)/\sinh(\Phi_0 - ik_{KH}L)\right]$$

$$(5.8.16)$$

with

$$\coth(\Phi_0) = z_r \qquad (5.8.17)$$

and v_L equal to the speed at $z = L$. The fluid speed at the opening, which determines the source strength S_ω, is therefore

$$v_0 = v_L \left[\sinh(\Phi_0)/\sinh(\Phi_0 - ik_{KH}L)\right]$$
$$= v_L/ \left[\cos(k_{KH}L) - iz_r \sin(k_{KH}L)\right] \qquad (5.8.18)$$

and the ratio v_0/v_L varies with $k_{KH}L$ is exactly the same way as $p_L/2p_i$ varies for the undriven tube subject to an external sound field (equation (5.8.15) and figure 5.18). When the waveguide is tuned so that $\text{Re}(k_{KH}L) = n\pi$ and the specific acoustic radiation impedance is not too large, v_0/v_L is at a minimum of just slightly less than unity; again, an acoustic image of the transducer is projected on to the opening of the waveguide. Usually $|z_r|$ is considerably less than unity and this characteristic behaviour therefore provides a means of tuning the waveguide by locating the length corresponding to a minimum in the transmitted amplitude.

If the external medium is unbounded then all of the energy radiates outwards and there are no reflected waves returning to the opening. Under those conditions, and assuming that $kb \ll 1$, equation (5.2.29) provides a good estimate of z_r. However, when coupled to an acoustic cavity there will be reflected waves and the problem is slightly more complicated. Under the assumption that only one mode of the cavity is excited, the appropriate expression for z_r is equation (5.2.37) which includes the effect of the acoustic pressure associated with the reflected waves but ignores their effect on the fluid speed. This will be a satisfactory approximation, except near to a resonance of the waveguide, because the intensity of the outgoing waves must greatly exceed that of the returning waves. The waves reflected back on to the opening actually see a different impedance there, equal to that for the passive tube considered above. Consequently, the perturbations caused to the resonance frequencies and linewidths of the cavity are unaffected by the activity of the source, as one would expect.

A practical limitation of these waveguides is that they limit the active diameter of the transducer to the diameter of the tube. However, it is possible to introduce a flared section into the tube which, if it is sufficiently gradual, does not introduce spurious reflections. A plane wave traversing such a section maintains approximately constant energy flux across any plane perpendicular to the axis, so a transition from radius b_1 to radius b_2 modifies the amplitudes of both the acoustic pressure and the fluid velocity by the ratio $(b_1/b_2)^2$. The flare, then, acts as an acoustic transformer with 'turns' ratio equal to the ratio of the input to output surface areas. It could be used to couple a transducer of radius b_2 to a resonator through an opening of radius b_1. When the waveguide is tuned for $\text{Re}(k_{KH}L) = n\pi$, a transformed image of the transducer is projected on to the opening at

$z = L$. If the transducer is the source then the fluid velocity is stepped up by the transformation (radius $b_2 > b_1$) although the effective strength, equal to the product of the area and the fluid velocity, is unaffected; but if the transducer is operated as a detector then the acoustic pressure reaching its surface is reduced by the transformation. Finally, since a wave entering the waveguide and being reflected at the far end back towards the opening returns unaffected by the presence of the flare, the admittance of the opening, and hence the perturbation caused by imperfect tuning, is also unaffected by the transformer (except through the dissipation, assumed small).

5.8.4   The open tube

The case of a tube open to free space may also be of interest. To a first approximation the acoustic pressure at the termination will then be zero so that the specific acoustic *impedance* vanishes there: $\coth(\Phi_L) = 0$ and $\Phi_L = \pm i\pi/2$. Since $\tan(x \pm i\pi/2) = \coth(x)$, this approximation gives

$$y_0 = \coth(ik_{KH}L) = -i \cot(k_{KH}L). \qquad (5.8.19)$$

When the dissipation is small, $|y_0|$ will again exhibit sharp minima and maxima but, in this case, the minima occur for $\mathrm{Re}(k_{KH}L) = (n + \frac{1}{2})\pi$ and the maxima for $\mathrm{Re}(k_{KH}L) = n\pi$. The terminal impedance actually differs from zero because of the effects of radiation from the open end of the tube at $z = L$; the boundary condition there should be written $\coth(\Phi_L) = z_r$, where z_r is the specific acoustic radiation impedance. When the frequency is sufficiently small, z_r is small in comparison with unity and the parameter Φ_L is approximately $z_r \pm i\pi/2$. The input specific acoustic admittance of the tube becomes

$$y_0 = \coth(ik_{KH}L + z_r) = -i \cot(k_{KH}L - iz_r). \qquad (5.8.20)$$

Under such low-frequency conditions, the fluid in the tube acts like a piston source and, since kb is also small compared with unity, the simple approximations of §5.2.5 give the radiation impedance of flanged or unflanged open tubes (equations (5.2.29) and (5.2.30)). Since in leading order iz_r is purely real, one can account for the radiation from the open end using an effective length L' such that $y_0 = -i \cot(k_{KH}L')$, where the 'end correction' $(L' - L)$ is equal to $(8/3\pi)b$ and about $0.6b$ respectively for flanged and unflanged tubes.

The main use of equation (5.8.20) would appear to be when calculating the perturbation caused by a simple tubular opening in the wall of a resonator, such as one through which the fluid is admitted. When used to couple sound into or out of a resonator, a waveguide is usually either closed by the face of the transducer or else terminated in a small chamber containing the transducer. In the latter case, when the internal dimensions of the volume are small compared with the wavelength of the sound, the terminal impedance may be estimated from a lumped circuit model which treats the chamber as an acoustic capacitor. The fluid in the opening of the tube at $z = L$ is treated like a massless piston moving with speed $v_L \exp(i\omega t)$, alternately compressing and expanding the fluid in the chamber. This process is assumed to take place isentropically so that the acoustic pressure in the chamber is $-\delta V/V\kappa_S$, where V is the volume of the chamber, $\delta V \exp(i\omega t)$ is the volume swept by the 'piston' and κ_S is the isentropic compressibility. Since $\delta V = i\omega v_L S$ and $u^2 = 1/\rho\kappa_S$, this model gives

$$z_L = -iS/kV \tag{5.8.21}$$

for the specific acoustic impedance of the termination where S is the cross-sectional area of the tube. Thus $\tan(i\Phi_L) = -(kV/S)$ and hence

$$y_0 = i[\tan(k_{KH}L) + (kV/S)]/[1 - (kV/S)\tan(k_{KH}L)]. \tag{5.8.22}$$

When z_L is small compared with unity, this y_0 reduces to $-i\cot(k_{KH}L - iz_L)$ and the system is similar to an open tube. However, the combined tube–chamber system exhibits an additional low-frequency resonance at which $|y_0|$ becomes large; this corresponds to the lowest root of $\text{Re}[\tan(k_{KH}L)] = S/kV$. Essentially, terminating the tube in a finite volume has the effect of raising the zero-frequency resonance of y_0 exhibited by the open tube to a finite value. This is illustrated in figure 5.20, which shows y_0 for two cases involving the 1 mm × 20 mm waveguide considered previously. In the first case (full line), the tube is open at $z = L$ (and z_L is taken as zero) while in the second, the tube is terminated by a closed chamber of volume 30 mm^3. If one takes the limit $V \to 0$ then equation (5.8.14) reduces to the expression for the rigidly closed tube.

When $\alpha_{KH}L \ll kL \ll 1$ near the low-frequency resonance of a tube–chamber system, the dissipation affects only the damping of the resonance, $\tan(k_{KH}L) \approx kL$; the angular frequency ω_0 is then given approximately by the familiar expression for the Helmholtz resonator:

$$\omega_0 = u(S/LV)^{1/2}. \tag{5.8.23}$$

In this approximation, the tube is equivalent to a lumped circuit element of mass ρLS (an approximation that is valid when $L \ll \lambda$).

The main disadvantage of this kind of arrangement is the low coupling efficiency. Since $|z_L|$ is small, the acoustic pressure generated inside the chamber by an external sound field is small compared with that at the opening (unless the tube is resonant). In fact, when S/kV is small, the two are related by

$$p_L \approx p_0/[\cos(k_{KH}L) - (kV/S)\sin(k_{KH}L)]. \tag{5.8.24}$$

At the frequencies where $|y_0|$ is minimum, so too is the acoustic pressure

at the transducer: $\mathrm{Re}(k_{KH}L) \approx (n + \frac{1}{2})\pi$ and $p_L \approx - [SL/(n + \frac{1}{2})\pi V] p_0$. [15]
Similarly, when the transducer inside the chamber is the driving source, the fluid speed at the opening (and hence the effective source strength) is a minimum when $|y_0|$ is minimum. In both cases, the efficiency near to $\mathrm{Re}(k_{KH}L) = (n + \frac{1}{2})\pi$ drops off rapidly at frequencies above the Helmholtz resonance. Increased efficiency can only be achieved by operation at intermediate frequencies where the waveguide will cause an increased perturbation. Often this can be tolerated provided that the resonance conditions of the tube are not approached too closely.

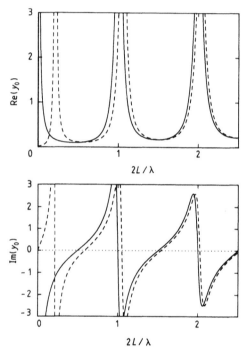

Figure 5.20 Input specific acoustic admittance y_0 of a cylindrical waveguide of length $L = 20$ mm and diameter 1 mm as a function of $\mathrm{Re}(k_{KH}L/\pi) = 2L/\lambda$: —————, end open to free space; – – – –, terminated by a closed chamber of volume 30 mm^3. The medium is argon at 300 K and 0.1 MPa.

References

[1] Morse P M and Ingard K U 1968 *Theoretical Acoustics* (New York: McGraw-Hill) pp 383–4
[2] Kinsler L E, Frey A R, Coppens A B and Sanders J V 1982 *Fundamentals of Acoustics* 3rd edn (New York: Wiley) p 202

[3] Merhaut J 1981 *Theory of Electroacoustics* English edn (New York: McGraw-Hill) pp 71–2

[4] Rasmussen G 1963 *Bruel and Kjaer Technical Review* No. 1

[5] Zuckerwar A J 1978 *J. Acoust. Soc. Am.* **64** 1278–85

[6] Zahn R 1981 *J. Acoust. Soc. Am.* **69** 1200–3

[7] Morse P M and Ingard K U 1968 *Theoretical Acoustics* (New York: McGraw-Hill) p 217

[8] Colclough A R 1979 *Acustica* **42** 28–36

[9] Morse P M and Ingard K U 1968 *Theoretical Acoustics* (New York: McGraw-Hill) pp 847–53

[10] Nolting B E and Terry N B 1957 *Technical Aspects of Sound* vol. 2 ed. E G Richardson (Amsterdam: Elsevier) ch. 2

[11] Berlincourt D A, Curran D R and Jaffe H 1964 *Physical Acoustics* vol. 1, part A, ed. W P Mason (New York: Academic Press) pp 202–4

[12] Manufacturer's data, Pennwalt Corporation, P O Box 799, Valley Forge, PA 19482, USA

[13] Van Randeraat J and Setterington R E (eds) 1974 *Piezoelectric Ceramics* 2nd edn (London: Mullard) ch. 4

[14] Van Randeraat S and Setterington R E (eds) 1974 *Piezoelectric Ceramics* 2nd edn (London: Mullard) ch. 6

[15] Manufacturer's data (Bruel & Kjaer microphone type 4134), Bruel & Kjaer, DK-2850 Naerum, Denmark

[16] Carey C, Bradshaw J, Lin E and Carnevale E H 1974 *Experimental Determination of Gas Properties at High Temperatures and/or High Pressures* (Arnold Engineering Development Center, Arnold Air Force Station, TN 37389, USA) *Report no.* AEDC-TR-74-33, also available as NTIS AD-779772

[17] Van Turnhout J 1980 *Electrets* ed. G M Sessler (New York: Springer) ch. 3

[18] Karbach A and Hess P 1985 *J. Chem. Phys.* **58** 3851–5

[19] Horowitz P W and Hill W 1980 *The Art of Electronics* 1st edn (Cambridge: Cambridge University Press) p 250

[20] Greenspan M and Thompson M C Jr 1953 *J. Acoust. Soc. Am.* **25** 92

[21] Colgate S O, Sona C F, Reed K R and Sivaraman A 1990 *J Chem. Eng. Data* **35** 1

[22] Kuhl W, Schodder G R and Schroder K K 1954 *Acustica* **4** 519

[23] Garland C W and Williams R D 1974 *Phys. Rev.* A **10** 1328

[24] Ewing M B, McGlashan M L and Trusler J P M 1985 *J. Chem. Thermodyn.* **17** 549

General references

In addition to those cited above, many of the references in Chapters 6 and 7 give useful information on the design of practical transducers. For quartz expander plates, references 9–11, 16, 17, 47 and 48 of Chapter 6 and reference 14 of Chapter 7 give design information. Solid dielectric capacitance transducers are described in references 25, 35, 38 and 39 of Chapter 6 and references 4 and 5 of Chapter 7. Information on moving-coil designs can be found in references 5, 6, 8 and 13 of Chapter 6.

Notes

1. This is the appropriate solution of the wave equation, $\nabla^2 p_a = -k^2 p_a$, in spherical coordinates.
2. The expression given in [7]; the final line of their equation (5.3.10) is too great by a factor of four.
3. This is not the only kind of solution allowed; for example, non-linearity can lead to the generation of subharmonics.
4. In the Bravais–Miller notation, the C_2 axes are labelled a while the C_3 axis is labelled c.
5. Lead zirconate titanate (PZT) ceramics are manufactured from mixtures containing $PbTiO_3$ and $ZrTiO_3$ as the main constituents.
6. The conventional notation of three-dimensional statics is used in this discussion of piezoelectricity with c for the matrix of elastic moduli and s for the compliance matrix. Elsewhere, s is used for stiffness in one-dimensional problems.
7. The matrix of elastic constants has diagonal symmetry and can therefore contain, at most, 21 unique elements.
8. The constant-E sound speed is appropriate for cases in which the strain and the electric field are perpendicular.
9. In principle, the frequency should be one of the parallel resonance frequencies so that, in addition to $\xi_0 \sin(kl)$, the term in the mechanical admittance containing the factor $\xi_1 \cos(kl)$ also vanishes. In practice, high mechanical impedance can also be achieved by inductive tuning at nearby frequencies.
10. The actual RMS noise voltage in bandwidth Δf is given by $e_n(\Delta f)^{1/2}$.
11. For a perfect gas $\kappa_T = 1/p$.
12. Solid rods have also been used [20].
13. This follows from the definition of the hyperbolic tangent: $\tanh(x) = [1 - \exp(-2x)]/[1 + \exp(-2x)]$.
14. $\alpha_{KH}L \approx 0.05(f/kHz)^{1/2}$ with $b = 0.5$ mm and $L = 20$ mm.
15. These approximations may be poor for $n = 1$ because of the effect of the Helmholtz resonance.

6

Experimental Methods 1:
Steady-state Techniques

6.1 Introduction

This chapter deals with the principal experimental techniques involving the use of a constant wave source. These methods generally depend upon standing waves being formed inside an acoustic cavity and they are therefore effective when the walls of the enclosure reflect sound efficiently. This situation is readily achieved when the fluid is a gas but for liquids the characteristic acoustic impedance is a significant fraction of that for a typical wall material and the reflection coefficient can be much lower. Consequently, the steady-state methods have been most successful with gases up to pressures of order 10 MPa or so.

In choosing the method best suited for a particular application, a number of trade-offs inevitably present themselves to the experimenter. In the present context these revolve mainly around two choices: (i) selection of either a fixed- or a variable-volume acoustic cavity; and (ii) operation of the cavity at either low frequencies or high frequencies. Variable-volume methods involve apparatus of considerably greater complexity but absolute values of the sound speed can be obtained without absolute measurement of length or volume; only displacement is required. Low-frequency operation, where the wavelength of the sound is of the same order as the dimensions of the cavity, usually enables a single normal mode to be studied in isolation without overlap with adjacent modes. However, the boundary layer effects may then be quite large and in some cases difficult to calculate with sufficient accuracy. High-frequency methods, on the other hand, avoid large boundary layer corrections at the expense of loss of resolution between modes.

In essence, the variable pathlength interferometer consists of a hollow tube fitted with a movable piston and closed at one end by the active surface of a transducer. The transducer is operated at a fixed frequency

179

producing what are assumed to be plane waves and the cavity is tuned through successive longitudinal resonances by moving the piston. When the reflection coefficients at the two ends of the interferometer are both unity, these resonances occur when the length is adjusted to an integer multiple of ($\lambda/2$), where λ is the wavelength of sound in the fluid under the operating conditions. Resonance can be detected in one of two ways. First, the electrical or mechanical impedance of the loaded transducer may be measured as the piston is moved and used to infer the input mechanical impedance of the fluid column. Alternatively, the reflector may be fitted with a second transducer used to monitor the acoustic pressure there as a function of displacement. This leads to a classification of interferometers as either single- or dual-transducer systems; we shall discuss each in turn. If wide-band transducers are available then the second method can be applied with variable frequency at constant pathlength and we have a simple fixed-volume resonator. The fixed-geometry systems are best classified by shape; both cylindrical and spherical designs will be discussed.

6.2 The Single-transducer Interferometer

6.2.1 Impedance and admittance figures

The basic working equation of the single-transducer variable pathlength interferometer, which gives the input mechanical impedance of the fluid column, is due to the labours of Hubbard [1, 2] as refined by Colclough [3]. It may be obtained from equation (3.8.4) which gives

$$Z_m = \mathcal{Q}p_a(z=0)/v_t = \mathcal{Q}\left(\frac{\omega\rho}{k_{01}}\right)\chi_t^{1/2}\left(\frac{\exp(ik_{01}L) + \chi_r \exp(-ik_{01}L)}{\exp(ik_{01}L) - \chi_r\chi_t \exp(-ik_{01}L)}\right) \quad (6.2.1)$$

where $\mathcal{Q}$ is the area of the source transducer, L is the length of the cavity, and χ_r and χ_t are the (complex) reflection coefficients at the reflector and at the transducer. As before, k_{01} is the propagation constant for the plane-wave transmission mode which we now write as

$$k_{01} = (\omega/u_{01}) - i\alpha_{01}. \quad (6.2.2)$$

In equation (6.2.2), u_{01} is the phase speed for plane waves travelling along the tube, $\alpha_{01} = \alpha + \alpha_{KH}$ is the corresponding absorption coefficient, and α_{KH} is the Kirchhoff–Helmholtz tube attenuation coefficient given by (2.4.34). To obtain the speed of sound u in free space, u_{01} must be corrected for the viscothermal effects at the wall according to equation (2.4.35) which gives

$$u = u_{01} + (u_{01}^2/\omega)\alpha_{KH} \quad (6.2.3)$$

correct to first order. The experimentalist is offered a choice here: the value of α_{KH} required to correct the measured phase speed can either be calculated from knowledge of the transport properties of the fluid or be obtained from the measured absorption coefficient after subtraction of the (usually much smaller) bulk absorption.

To facilitate the discussion we shall begin by assuming that the reflection coefficients are both unity so that the only dissipation comes from the transmission losses along the tube. In that case, and setting the prefactor $(\omega\rho/k_{01})$ equal to ρu, the input mechanical impedance is given by the simple equation

$$Z_m/\mathcal{C}\rho u = -\mathrm{i}\,\cot(k_{01}L). \tag{6.2.4}$$

The resonances and anti-resonances of the fluid column are defined by the vanishing of the imaginary part of Z_m. The resonances occur at $\mathrm{Re}(k_{01}L) = n\pi$ and correspond to maxima of $\mathrm{Re}(Z_m)$, while the anti-resonances occur at $\mathrm{Re}(k_{01}L) = (n+\frac{1}{2})\pi$ and correspond to minima of $\mathrm{Re}(Z_m)$.[1] The lengths L_n^r and L_n^a at which the cavity is tuned to resonance and anti-resonance respectively are therefore given by

$$L_n^r = n\pi(u_{01}/\omega) \tag{6.2.5a}$$

$$L_n^a = (n+\tfrac{1}{2})\pi(u_{01}/\omega). \tag{6.2.5b}$$

In figure 6.1, the behaviour of Z_m is shown (with arrows indicating the direction of increasing pathlength) as the cavity is tuned through the first few resonances. When, as illustrated, the absorption over one pathlength is small, the reduced impedance $Z_m/\mathcal{C}\rho u$ is equal to the small quantity $\alpha_{01}L_n^a$ at anti-resonance and the large quantity $1/\alpha_{01}L_n^r$ at resonance. To the extent that the impedance at anti-resonance is negligible, Z_m takes the form of a series of circles lying symmetrically about the real axis, touching the origin, and with diameters

$$D_n^r = \mathcal{C}\rho u/\alpha_{01}L_n^r \tag{6.2.6}$$

which stand in the ratio $1:1/2:1/3:\cdots:1/n:\cdots$. At a closer level of scrutiny, these circles are actually the loops of a single spiral which eventually converges on the point $Z_m/\mathcal{C}\rho u = 1$ when $L \to \infty$ and the interferometer becomes an infinite transmission line. Consequently, the interferometer provides useful information only in the region of high standing wave ratio.

Close to each resonance, $\tan(k_{01}L)$ may be expanded as a series of odd powers of $(k_{01}L - n\pi)$ the first of which has the coefficient unity. Consequently, when $\alpha_{01}L \ll 1$, the impedance has the behaviour of a Lorentzian function near to resonance:

$$Z_m = \frac{\mathcal{C}(\rho u^2/\omega)}{\Delta L_n^r + \mathrm{i}(L - L_n^r)} \tag{6.2.7}$$

with the half-width ΔL_n^r given by

$$\Delta L_n^r = (u_{01}/\omega)\alpha_{01}L_n^r. \tag{6.2.8}$$

The portion of the impedance circle between the points at which the phase angle ϕ equals $\pm\pi/4$ is traversed rapidly during the small displacement of the piston from $L=(L_n^r-\Delta L_n^r)$ to $L=(L_n^r+\Delta L_n^r)$. Clearly, in our present approximation the sound speed can be obtained from a measurement of the absolute length of the cavity at resonance (where $\phi=0$) and the absorption coefficient can be obtained from the displacement required to rotate the phase between the points $\phi=\pm\pi/4$.[2] In the past, this information has been obtained graphically by actually plotting the impedance circle [4–6]. With modern laboratory equipment the data are likely to be logged by a computer which can then be used for rapid numerical analysis of the data to obtain this information. Often, phase-sensitive detection has not been employed and only the modulus $|Z_m|$ has been measured. This takes the form, illustrated in figure 6.1(b), of a symmetric resonance curve with its maximum at $L=L_n^r$ and with half-width ΔL_n^r at $1/\sqrt{2}$ of the maximum height. Again, either graphical or numerical analysis of the data is possible.

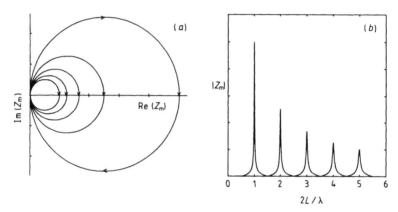

Figure 6.1 Reduced input mechanical impedance $Z/\mathbb{Q}\rho u$ for an interferometer operating with $\alpha_{01}\lambda_{01}=0.02$: (a) impedance circles for $n=1$ to 5; (b) $|Z_m/\mathbb{Q}\rho u|$ as a function of $2L/\lambda_{01}$.

In our present approximation, neglecting losses and phase shifts on reflection, the input mechanical admittance $Y_m=Z_m^{-1}$ is given by

$$\mathbb{Q}\rho u Y_m = i\,\tan(k_{01}L) \tag{6.2.9}$$

which, like Z_m, also describes a series of circles in the complex plane when the standing wave ratio is high. In the absence of a contribution from the intrinsic impedance of the transducer itself, the signal-to-noise ratio for admittance measurements would be most favourable near anti-resonance where $|Y_m|$ is maximum and another Lorentzian-type function applies:

$$Y_m = \frac{\omega/\alpha\rho u^2}{\Delta L_n^a + i(L - L_n^a)}. \tag{6.2.10}$$

Now the phase of the measured quantity rotates between $\pi/4$ and $-\pi/4$ as the length is increased between $L_n^a - \Delta L_n^a$ and $L_n^a + \Delta L_n^a$, where ΔL_n^a is the anti-resonance half-width given by

$$\Delta L_n^a = (u_{01}/\omega)\alpha_{01} L_n^a. \tag{6.2.11}$$

6.2.2 End-face corrections

Before going any further we must restore the effects of loss and phase shift on reflection. To do that, one returns to equation (6.2.1) and solves for the lengths at which $\tan(\phi)$ vanishes; actually a solution accurate just to first order in the small quantities $(1 - \chi_r)$ and $(1 - \chi_t)$ will be sufficient. In that (very good) approximation, the impedance and admittance figures remain circular but the lengths, half-widths and circle diameters all change slightly. In first order, the resonance condition corresponds to the vanishing of the imaginary part of the denominator and therefore involves only the product of the two end-face reflection coefficients. Consequently, it is convenient to write the geometric mean of the two coefficients as

$$(\chi_r\chi_t)^{1/2} = \exp(i\delta - \beta) \tag{6.2.12}$$

where δ is a phase shift and β is an absorption coefficient. Then, to first order in the small quantities $\alpha_{01} L$, β and δ, the resonance length, half-width and impedance circle diameter are given by

$$L_n^r = (n\pi + \delta)(u_{01}/\omega) \tag{6.2.13}$$

$$\Delta L_n^r = (\alpha_{01} L_n^r + \beta)(u_{01}/\omega) \tag{6.2.14}$$

$$D_n = \alpha\rho u/(\alpha_{01} L_n^r + \beta). \tag{6.2.15}$$

These are exactly the results that follow from intuitive arguments regarding the phase shift and absorption on reflection. If the sound speed is obtained from a measurement of the displacement between two resonances then no correction for the phase shift on reflection arises. However, if the absolute length is used then a calculated correction for δ may be required. Both transmission and end-face absorption coefficients can be obtained from the measured half-widths and circle diameters, so that a measured correction of the sound speeds for the boundary layer effects at the sides of the interferometer can be applied. At low frequencies in gases, the reflection coefficients differ from unity primarily because of the thermal boundary layers, so that $\chi_r \approx \chi_t$ and

$$\beta - i\delta \approx 2y_h. \tag{6.2.16}$$

An indication of the order of magnitude involved is afforded by the values $\beta = -\delta \approx 1 \times 10^{-3}$ in an interferometer operating at 1 kHz and filled with argon gas at 300 K and 0.1 MPa. Second-order terms, which involve cross-products of $\alpha_{01}L_n^r$, β and δ, are typically 1×10^{-5} or less and almost certainly no larger than the imprecision in experimental measurements of the resonance lengths. This justifies the adoption of a first-order treatment of end-face corrections. In dense fluids, or at frequencies near to a resonance of the end plate, mechanical terms may also be important in determining the reflection coefficient χ_r. However, because of the way in which Z_m is defined, mechanical losses do not contribute to the reflection coefficient at the face of the transducer (see §3.8.3).

At anti-resonance, first-order expressions are obtained by requiring the imaginary part of the numerator of equation (6.2.1) to vanish. This involves only the reflection coefficient χ_r which may be written as $\exp(i\delta_r - \beta_r)$, where both β_r and δ_r are of the first order of smallness. The anti-resonance length, half-width and the diameter of the nth admittance circle then become

$$L_n^a = [(n + \tfrac{1}{2})\pi + \tfrac{1}{2}\delta_r]\,(u_{01}/\omega) \tag{6.2.17}$$

$$\Delta L_n^a = (\alpha_{01}L_n^a + \tfrac{1}{2}\beta_r)(u_{01}/\omega) \tag{6.2.18}$$

$$D_n^a = [a\rho u(\alpha_{01}L_n^a + \tfrac{1}{2}\beta_r)]^{-1}. \tag{6.2.19}$$

The approximation $(\omega\rho/k_{01})\chi_t^{1/2} = \rho u$, used above, leads to further corrections but they are of the second order of smallness [3] and will be considered no further here.

6.2.3 Non-zero transducer impedance

Equations (6.2.3) and (6.2.9) assume that we can somehow obtain a measure of the input mechanical impedance or admittance of the fluid column unburdened by any other contribution. In practice the driving force is applied to the moving element of a transducer and one can measure not Z_m alone but the sum of the intrinsic impedance of the transducer Z_t and the impedance of the fluid column. The effect of this additional impedance is, as shown in figure 6.2, to shift the origin of the impedance figures to $Z = Z_t$. The line joining the points of resonance and anti-resonance remains parallel to the real axis and it is possible to analyse the data graphically to obtain the length of the cavity at resonance and the resonance half-width as before. Although requiring non-linear methods, numerical analysis of the data remains straightforward; one simply includes a complex constant in the fit to account for the intrinsic impedance of the transducer. The use of phase-sensitive detection is particularly advantageous because it provides more information against which to optimize the parameters in the analysis (be that graphical or numerical). However, it remains possible in principle to obtain all the required information from just $|Z|$ as a function of L. Figure 6.3 shows the modulus of the combined mechanical impedance of the transducer and the fluid column for a case in which both real and imaginary parts of Z_t are substantial. Numerical analysis of the data will be the usual approach. However, one can construct the impedance circle graphically from the $(|Z|, L)$ data [3]. The diameter D_n^r of each impedance circle may be obtained from the difference between the local maxima and minima of $|Z|$ while the mean of those two values will be the distance of the centre of the circle from the origin in the complex plane. Furthermore, the point of anti-resonance must lie at a distance $|Z_t|$ from the origin and, since $|Z_t|$ may be equated with the value of $|Z|$ far from resonance, the circle can be located in the complex plane and the resonance parameters determined. This of course assumes that the impedance data actually do form a circle as predicted by the theory— an assumption that can only be tested using both modulus and phase information.

In the case of admittance data, the effect of the intrinsic impedance of the transducer can be radical. This is because, with the possible exception of a resonant piezoelectric transducer, the intrinsic impedance will greatly exceed the impedance of the fluid column at anti-resonance. Consequently,

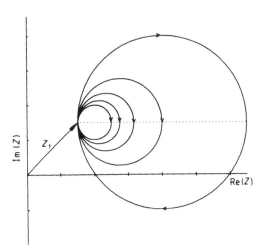

Figure 6.2 Reduced combined mechanical impedance $[(Z_m + Z_t)/Q\rho u]$ of fluid column and transducer plotted for $\alpha_{01}\lambda_{01} = 0.02$ and $Z_t = 0.3(1 + i)/\alpha_{01}L_1^r$. The dotted line shows the main resonance–anti-resonance axis.

the rapid variation of Y near to anti-resonance given by equation (6.2.9) will be masked by the transducer and a sensitive measurement of L_n^a will be impossible. This does not, however, render admittance data useless. If one calculates the magnitude of $(\mathrm{d}Y/\mathrm{d}L)$, which measures the sensitivity, one finds turning points at lengths near to both the resonance and the anti-resonance points. Differentiating again, one finds that, when $|Z_t|/\mathfrak{A}\rho u > 1$, the length $L_n^{(A)}$ at which Y is changing most rapidly now lies near to resonance and that minimum sensitivity is near to anti-resonance; only when $|Z_t|/\mathfrak{A}\rho u < 1$ is the reverse true. Consequently, for cases in which the transducer has a mechanical impedance large compared with the characteristic input impedance $\mathfrak{A}\rho u$, admittance data should be collected for lengths near to $L = L_n^{(A)}$ rather than near anti-resonance as suggested by equation (6.2.10). The inverse of course gives $Z_m + Z_t$ and numerical analysis can then be the same as with impedance data.

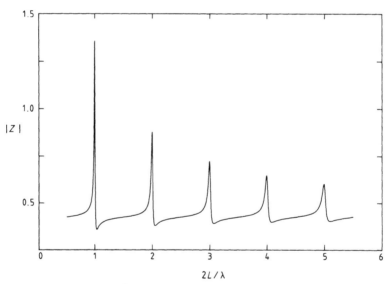

Figure 6.3　$[(Z_m + Z_t)/\mathfrak{A}\rho u]$ as a function of $2L/\lambda$. The conditions are the same as in figure 6.2.

One can show that if, at anti-resonance, Z_m is negligible compared with Z_t then the admittance function is, like the impedance function, a series of circles in the complex plane touching at the point of anti-resonance [7, 8]. This situation is illustrated in figure 6.4. It is shown in [7] using conformal mapping that the admittance circles share a common axis which passes through the position of anti-resonance and which can be used to define a phase $\phi^{(A)}$ relative to an origin at $Y = Z_T^{-1}$. The point on each admittance circle opposite to the position of anti-resonance, at which $L = L_n^{(A)}$, does not in fact correspond exactly to resonance but it can be shown to have a constant phase relation with the position of resonance (independent of n) when we measure phase in terms of the angle $\phi^{(A)}$ [7, 8]. In terms of $Z_t = R_t + iX_t$, the length $L_n^{(A)}$ is

$$L_n^{(A)} = L_n^r + (\alpha \rho u^2/\omega)(X_t/|Z_t|^2) \qquad (6.2.20)$$

and the diameter of the admittance circle is given by

$$D_n^{(A)} = D_n^r/(|Z_t|^2 + R_t D_n^r). \qquad (6.2.21)$$

A half-width can also be defined by the displacement from $L_n^{(A)}$ necessary to rotate the phase $\phi^{(A)}$ by $\pm \pi/4$ and is given by

$$\Delta L_n^{(A)} = \Delta L_n^r + (\alpha \rho u^2/\omega)(R_t/|Z_t|^2). \qquad (6.2.22)$$

Consequently, the wavelength and the absorption coefficient α_{01} can be obtained from the admittance circles without error using differences in $L_n^{(A)}$ and $\Delta L_n^{(A)}$ between two orders of resonance. When the transducer impedance is fairly small, so that $L_n^{(A)}$ and L_n^r differ significantly compared with ΔL_n^r, the data are best acquired symmetrically about $L_n^{(A)}$ (rather than about L_n^r itself) and analysed on the basis of equations (6.2.20–22) so that the maximum sensitivity is utilized.

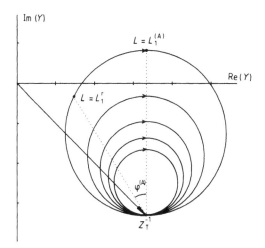

Figure 6.4 Combined mechanical admittance $[\alpha \rho u/(Z_m + Z_t)]$ of fluid column and transducer. The conditions are the same as in figure 6.2.

When a resonant piezoelectric transducer is used, its reactance X_t will be zero but R_t may not be negligible compared with the very small impedance of the fluid column at anti-resonance because, even with nodal mounting, the transducer is still subject to damping by the fluid at its rear face. If the transducer is in contact with the same medium on both sides then the smallest value that the resistance R_t could possibly take is one approximately *equal* to the input impedance of the fluid column at anti-resonance (this would require quarter-wave backing [9]). Only if the interferometric fluid is a liquid or a dense gas and the backing medium a dilute gas can admittance circles with maxima at anti-resonance be obtained in practice.

6.2.4   *Measurement of impedance or admittance*

In principle there is no reason to favour measurement of Z_m over that of Y_m or vice versa; the choice will be dictated by the available instruments. In practice we must infer the input impedance or admittance of the fluid column from the measured mechanical or electrical impedance of the transducer. The several possibilities may be typified by two practical cases. First, there is the familiar case of a thickness-mode piezoelectric transducer, usually of x-cut quartz, driven at one of its natural frequencies. Second, there are moving-coil-driven diaphragms, which have been used at audio frequencies under conditions where the intrinsic impedance is comparable with the input impedance of the interferometer.

When suitably mounted, the mechanical impedance of a resonant piezoelectric expander plate will be the sum of the mechanical impedance presented to the 'active' face of the transducer by the interferometer and a small load impedance due to radiation from the other face and/or losses in the mounting. When the standing wave ratio in the interferometer is high the unwanted loading can be made small compared with the load impedance of the interferometer at resonance. As discussed above, this cannot usually be achieved at anti-resonance; however, this of little consequence in resonance measurements. As described in Chapter 5, the effects of the static capacitance of the transducer can be tuned out with inductors (as can stray cable capacitance) so that the electrical and mechanical impedances are in direct proportion when the transducer is driven by an ideal voltage source. Under those conditions, one measures either the magnitude or the magnitude and phase of the current drawn by the transducer at constant excitation voltage as a function of piston displacement and consequently one obtains an analogue of mechanical admittance. If this is done using a shunt resistor then its value should be small in comparison with the electrical equivalent of the mechanical load resistance. Alternatively, the transducer may be driven by a constant-current source and the magnitude, or the magnitude and phase, of the potential difference developed across its terminals measured. This also provides a measure of the mechanical admittance, the magnitude of which is a minimum at the resonance of the interferometer. It is not necessary to measure the full impedance or admittance figure, only that portion of it near to resonance (say one or two half-widths on either side of L_n^r). Radio frequency impedance/admittance analysers, suitable for measurement of both magnitude and phase, are available commercially but, like dual-phase lock-in detectors for these frequencies, tend to be rather expensive. However, in the absence of a significant contribution from Z_t near to resonance, the response curve is simple and it is quite sufficient to measure just the magnitude of the impedance or admittance using a radio frequency voltmeter.

The preparation and mounting of quartz crystals, and the attachment of electrical connections, are matters for some experimentation [10, 11]. Unless one exercises care, true thickness-mode operation will not be achieved because of coupling with other modes of the crystal. Bevelling the edges of the crystal is said to reduce substantially the effects of unwanted face shear modes [12]. Ideally, the purity of the mode should be studied by measuring the input electrical impedance as a function of frequency in the vicinity of the crystal resonance frequency [10]. In principle, nodal mounting of x-cut expander plates [9] offers the lowest possible damping but it is difficult to realize with thin crystals. When the acoustic load impedance is going to be large (as, for example, in the study of gases at high pressures) then the damping in the mounting is less important and the crystal can be edge supported or mounted on a backing plate.

At low frequencies, piezoelectric expanders become too large to be useful and moving-coil-driven diaphragms have been used [5, 6, 8, 13]. Here the transducer may or may not be driven at a resonance but the damping is usually quite large so that, even at a resonance of the transducer, Z_t may not be negligible. The electrical input impedance can be measured as for the piezoelectric transducer but now, in addition to the unwanted intrinsic mechanical impedance, the measurement is further burdened by the intrinsic electrical impedance of the of the coil. One successful approach that has been used involves direct measurement of an analogue of the mechanical admittance of the diaphragm [8, 13]. In this method, the coil is driven at constant current, so that the diaphragm experiences a constant force amplitude, and an accelerometer attached to the rear face of the diaphragm is used to monitor the displacement. By measuring the output of the accelerometer using a two-phase lock-in detector, triggered by a reference voltage developed across a shunt resistor in series with the drive coil, both real and imaginary components of the mechanical admittance were obtained (in arbitrary units, of course, but this does not matter). In another approach, the diaphragm was driven at constant displacement amplitude (using an optical monitoring method) and a measure of impedance obtained from the required drive voltage [5, 6].

It may be attractive to use a source, such as a capacitance transducer, driven under non-resonant conditions where the input electrical impedance is very large and the relative variation resulting from the fluid load is quite small. An appropriate method then of measuring the small change in impedance or admittance accompanying displacement of the piston is a technique in which the transducer forms one element of an AC bridge and the out-of-balance signal is measured using a lock-in detector. Dual-phase lock-in amplifiers operating at frequencies up to about 100 kHz are available commercially offering good performance at modest cost.

6.2.5   A low-frequency design

The apparatus of Colclough *et al* [14] was designed for an acoustic measurement of the gas constant using argon gas as the working fluid at a temperature close to 273.16 K and at pressures between 0.03 and 1.3 MPa [8, 14]. A low-frequency (5.6 kHz) and a fairly narrow bore (30 mm) were chosen so that the instrument could be operated below the first cut-off frequency thus avoiding completely the possibility of unresolved higher modes. The penalty incurred by that approach was the requirement for rather large boundary layer corrections (about 0.16 per cent under typical conditions). Since the transport coefficients of the working gas were not known with sufficient accuracy to calculate the boundary layer corrections, the measured half-widths were used to obtain the required information.

The interferometer, shown in figure 6.5, consisted of a brass cavity fitted with a nickel–tin-plated brass piston which could be moved by means of a motorized lead screw assembly over a distance of about 150 mm (sufficient to cover five half wavelengths). The displacement of the piston was measured using an optical interferometer. The transducer was a moving-coil-driven aluminium alloy diaphragm, 0.25 mm thick, operated at constant current and fitted on its rear face with a PZT accelerometer for measurement of its displacement amplitude. Admittance circles were plotted directly on an x–y recorder using phase-sensitive detection of the output signal from the accelerometer as described above. The piston was stopped at three locations on each admittance circle, corresponding to phase angles $\phi^{(A)}$ near to zero and $\pm \pi/4$, for the length of the cavity and the phase of the accelerometer signal to be recorded for subsequent analysis of the circles. The wavelength was then obtained, with a typical imprecision in a single measurement of about 30 PPM (parts per million) at 0.1 MPa, from the displacement between the points $\phi^{(A)} = 0$ on the first and fifth admittance circles.

Unfortunately, the transducer proved to be slightly non-linear leading to a distortion in the shape of the admittance figures and systematic errors in the derived sound speeds [14, 15]. These errors were proportional to $\alpha_{01}\lambda_{01}$ and reached about 25 PPM at the lowest pressure. Once the problem was recognized, it proved possible to calculate corrections based on the ratio of the maximum and minimum 'diameters' of the distorted circles. The final value of the gas constant obtained from the corrected data was associated with a fractional imprecision of just 6 PPM (one standard deviation) and an equal level of systematic uncertainty. The reason for the non-linearity is not clearly understood, especially as the diaphragm was driven under non-resonant conditions where the displacement amplitude was very small; however, similar behaviour was found in a second apparatus constructed along the same lines and used at temperatures in the range 4.2–20 K [13]. It is tempting to speculate that the diaphragms were in radial compression as a result of edge clamping.

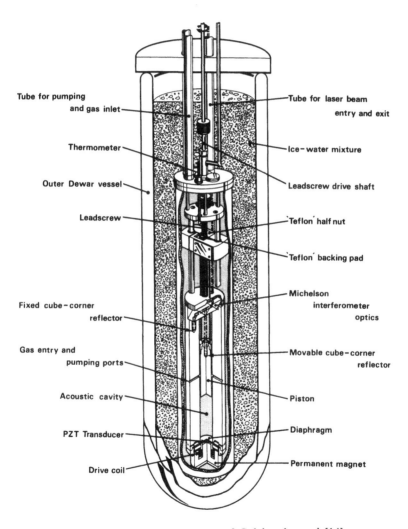

Figure 6.5 The interferometer of Colclough *et al* [14].

6.2.6 A high-frequency design

The classical interferometer operates with a quartz crystal transducer at high frequencies where the wavelength is quite small (e.g. 0.3 mm in argon at 1 MHz, 300 K and 0.1 MPa). Under these conditions, sufficient parallelism between end faces is much harder to achieve. Although the pure longitudinal modes are relatively insensitive to this geometric error, the frequency is generally well above the first cut-off value for higher modes and any tilting of the end faces will give rise to unwanted coupling with these modes. In severe cases 'satellite' resonances can be observed.

The apparatus used by Henderson and Peselnick [16] and illustrated in figure 6.6 will serve as an example. The piston and cylinder, which were of stainless steel, formed an acoustic cavity of diameter 25.4 mm with length variable up to about 75 mm by means of a micrometer lead screw. The interferometer was mounted inside a pressure vessel and thermostat. To ensure parallelism between transducer and reflector, the piston had been clamped in the cylinder at the position of zero displacement and the combined assembly ground flat and polished during manufacture to a tolerance better than one quarter wavelength of light. The *x*-cut quartz transducer,

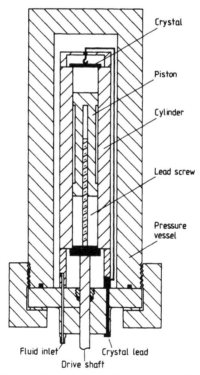

Figure 6.6 The interferometer of Henderson and Peselnick [16].

which was optically flat and gold plated, was then mounted over the open end of the cylinder and restrained only by a light spring used to make electrical contact with its rear face. This design does not provide the lowest possible residual mechanical impedance but it does provide the best chance of ensuring parallelism. Since the interferometer was used at high pressures (up to 23 MPa) where the acoustical load impedance was large, the residual impedance caused by the mounting was unimportant. By operating the transducer at harmonics of the fundamental, frequencies of between 0.3 and 7 MHz were obtained thereby allowing a comprehensive study of dispersion and absorption of sound in carbon dioxide right up to liquid densities.

6.2.7 Unguided waves

Some measurements of speed and absorption of sound in fluids have been performed using interferometers consisting just of a plane source and a parallel movable reflector without a waveguide. This approach has some merits because unwanted effects associated with the walls of the conventional interferometer, including both viscothermal and precondensation phenomena, vanish. As we have seen, phase shifts and losses on reflection can be made to cancel, so that a measurement free of viscothermal and precondensation errors is possible in principle.

The method is of course subject to diffraction errors because the source does not generate pure plane waves. As we have seen, the outgoing wave from a piston source exhibits a phase advance (relative to a plane wave) that asymptotically approaches $\pi/2$ at large separations. This diffraction phase advance leads to an apparent sound speed in excess of the free-space value. Similarly, beam spreading results in attenuation in excess of that for a plane wave. Actually, the effects of diffraction on interferometric measurements of the sound speed are very much smaller than one might expect on the basis of the free-field behaviour because of interference between positive- and negative-going waves which, near resonance, enhances the plane-wave component of the wave field. As with the so-called guided-mode dispersion,[3] diffraction effects decrease rapidly as the radii of the source and reflector are increased. But, as the absorption coefficient increases (and the standing wave ratio declines) the diffraction effects increase towards those of the free field.

The theoretical calculations of Krasnushkin [43], which assume perfect piston-like motion of the source and small losses, lead to analytic expressions for the apparent excess sound speed Δu and excess absorption coefficient $\Delta\alpha$:

$$\Delta u/u \approx 0.0317(\lambda^2/b^2) \qquad (6.2.23)$$

$$\Delta\alpha \approx 0.1\lambda/b^2. \qquad (6.2.24)$$

Detailed numerical calculations for a water-filled interferometer, including reflection losses, have been performed by Del Grosso [44, 45]. For a source and reflector of equal radius b, operating with separations in excess of 5λ, he found errors in the sound speed of less than 600 PPM at $b = 10\lambda$ falling to less than 30 PPM at $b = 50\lambda$. The errors became smaller still when the diameter of the reflector was made larger, although they increased as expected when absorption in the fluid was considered.

Such calculations have been used to optimize the design of unguided interferometers used to perform highly accurate measurements of the speed of sound in water at both ambient and elevated pressures [47, 48]. Pathlengths of order 300λ were used in these measurements so that the diffraction phase advance became a very small correction. A free-field method combining long pathlengths (up to 0.85 m) and theoretical diffraction corrections has also been exploited for the same purpose [49].[4] The results obtained for the speed of sound in pure water at atmospheric pressure by these two different experimental techniques are in remarkably good agreement: the maximum difference in the temperature range 276.5 to 307.0 K being just $0.07 \, \mathrm{m \, s^{-1}}$ (about 50 PPM) [49]. This suggests that diffraction errors can be rendered very small indeed by the use of long pathlengths and accurately aligned transducers.

For a fluid such as liquid water, in which the absorption coefficient is rather small at frequencies of order 1 MHz, the use of long pathlengths is a practical proposition. However, as discussed in connection with equation (5.2.16), the near-field radiation pattern approaches a beam of plane waves over a substantial portion of the source when $\exp(-\alpha b)$ is small. Consequently, operation of an unguided interferometer at small source–reflector separations ($L \ll b$) becomes attractive when the absorption in the fluid (or on reflection) is large [9].

6.3   The Double-transducer Interferometer

As a useful alternative to impedance measurements, a second transducer can be used to detect resonance by monitoring the acoustic pressure at the end of the interferometer remote from the source. In the regime of high standing wave ratio, the amplitude of the acoustic pressure there will vary with piston displacement showing sharp maxima separated by integer numbers of half wavelengths. Phase shifts on reflection are of no consequence when sound speeds are determined from the displacement between maxima of the received signal. However, the absolute lengths corresponding to maxima in the received signal will depend upon the boundary conditions at either end of the interferometer.

6.3.1 Boundary conditions

The acoustic pressure at the end remote from the source is given by equation (3.8.4). Let us first suppose that the reflection coefficients are both unity so that the acoustic pressure at $z = L$ is given by

$$p_a(z = L) = \omega \xi_t \exp(i\omega t)\rho u / \sin(k_{01} L) \qquad (6.3.1)$$

where ξ_t is the displacement amplitude of the source transducer. Provided that ξ_t is constant, the magnitude of the acoustic pressure at the detector has sharp maxima near to resonance and broad minima near to anti-resonance. In that case, and close to resonance, equation (6.3.1) becomes a Lorentzian function

$$p_a(z = L) = \frac{(-1)^n i \rho u^2 \xi_t}{\Delta L_n^r + i(L - L_n^r)} \qquad (6.3.2)$$

which varies as a function of pathlength in exactly the same way as does the input mechanical impedance. Note that constancy of ξ_t, despite variation in the load impedance, implies a source of infinite mechanical impedance. This situation may be approached in practice with a 'stiff' source such as a solid dielectric capacitance transducer with smooth back plate. When the reflection coefficients differ slightly from unity, the pathlength at resonance L_n^r and the half-width ΔL_n^r will be given by equations (6.2.13) and (6.2.14) and, if the absolute length of the cavity is to be used, the phase shift δ must be known. In practice, while the boundary layer contribution may be calculated, mechanical terms can usually only be estimated.

The situation with a resonant piezoelectric transducer as the source could be radically different. For example, in the extreme where the mechanical impedance of the source is negligible compared with that of the fluid column, the velocity $v_t = i\omega \xi_t$ at constant driving force F becomes F/Z_m, where Z_m is the input mechanical impedance of the fluid column. Since $Z_m \sim \cot(kL)$, the magnitude of the acoustic pressure at the detector would then have sharp maxima near $L = (n + \frac{1}{2})(\lambda/2)$ and broad minima near $L = n(\lambda/2)$. However, for the important case in which a parallel inductor is used to annul the static capacitance of the source transducer plus the cable at the open-circuit mechanical resonance frequency, and the network is driven at constant current, ξ_t will be sensibly constant. Furthermore, when the same inductive tuning is used for the detector crystal, and the electrical output is measured with a high input impedance detector, the reflection coefficient at $z = L$ should differ from unity only on account of the thermal boundary layer.

6.3.2 A high-frequency design

The interferometer of Gammon and Douslin [17, 18], shown in figure 6.7, is an example of a double-transducer apparatus for sound speed and absorption measurements at relatively high frequencies. This employed a quite large-bore (51 mm) cylinder within which a close-fitting piston, keyed

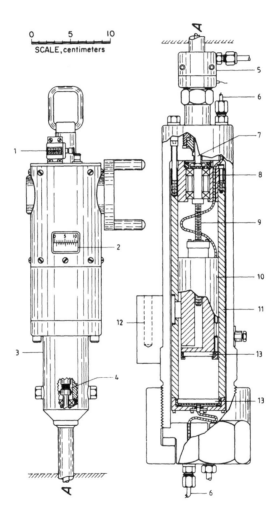

Figure 6.7 The interferometer of Gammon and Douslin [17, 18]: 1, mechanical counter; 2, vernier; 3, packing gland; 4, bellows coupling; 5, tempering block; 6, crystal lead; 7, drive shaft; 8, bearing assembly; 9, micrometer screw; 10, piston; 11, cylinder; 12, thermometer well; 13, x-cut quartz crystals.

to prevent rotation, could be moved. Both piston and cylinder were fabricated from a heat-treatable stainless steel and, in an attempt to avoid galling, the cylinder was annealed while the piston was hardened. On its lower face, the piston carried a 500 kHz quartz crystal transducer, while the base of the cylinder was closed by a second identical transducer. In order to achieve parallelism of the transducers, the crystal mounting surfaces were lapped at zero displacement during manufacture and the piston was spring loaded against one side of the bore. A micrometer screw, calibrated against gauge blocks, was used for changing and measuring the separation of the two transducers.

The source transducer was driven by an oscillator, the frequency of which was adjusted to an odd harmonic of the fundamental crystal resonance and measured with an electronic counter, while the output of the detector was measured with a wide-band voltmeter. Variable inductors were used to tune out capacitive reactances parallel to both transducers.

The interferometer was mounted inside a pressure vessel and thermostat for operation at pressures up to 25 MPa and temperatures up to 420 K. Typically, the intercrystalline separation at which resonance lengths were measured was between 50 and 100 mm and, by using the absolute length in the calculations, the reading precision and accuracy of the micrometer (respectively 0.3 and 1.3 μm) were used to their best advantage. Small corrections, based on Del Grosso's theoretical calculations [44], were applied for guided-mode dispersion. Comprehensive measurements in helium in the range 98–423 K and 1–15 MPa yielded a precision in the average sound speed from 12 resonances at each state point of order 10 PPM (one standard deviation) [17, 18]. The apparatus was also used to study methane on both sides of the gas–liquid phase boundary [19]. In the region of high absorption where the standing wave ratio was small (e.g. near to the critical point), the sound speed could still be obtained from the displacement required to advance or retard the phase of the received signal by $\pi/2$.

6.4 Spherical Resonators

Fixed-volume acoustic resonators are advantageous for low-frequency measurements of sound speed and absorption because, in that regime, the normal modes of the cavity are usually resolved one from another, and wide-bandwidth transducers are available with which to excite and detect them. The spherical geometry has proved to be ideally suited to highly precise measurements in gases under conditions of relatively small sound absorption.

6.4.1 Merits of the spherical geometry

The merits of the spherical geometry have been demonstrated by Moldover, Mehl and their collaborators in a series of papers developing both the theory and practice of the method and culminating in the most accurate determination yet of the universal gas constant [20–28].

Essentially, the principal advantages of the sphere lie in the existence of the radially symmetric modes which are characterized both by the absence of viscous damping at the surface and by resonance frequencies that are insensitive to geometric imperfections. The absence of viscous damping, and the favourable volume-to-surface-area ratio in the sphere, leads to higher quality factors in gases than for any other geometry of similar volume and operating frequency, and this in turn leads to the highest possible precision in the determination of the resonance frequencies. As an example, let us compare the quality factors Q of the lowest radial mode of a spherical and a cylindrical resonator chosen such that both have the same volume and the same resonance frequency of 5 kHz when filled with argon gas at 300 K and 0.1 MPa. Under these conditions, boundary layer losses dominate in both resonators and in the sphere ($a = 46$ mm) Q is 1880, while in the cylinder ($L = 85$ mm, $b = 39$ mm) Q is only 790. Had we chosen the lowest longitudinal mode of a cylinder for comparison under the same conditions, the Q would have been even lower at just 420. In polyatomic gases, the factor $\gamma - 1$ in the thermal boundary layer term is much smaller than for monatomic systems and, when bulk absorption is slight, the quality factors of the radial modes in the sphere can then be an order of magnitude greater still. Since, for gases at low pressures, boundary layer effects lead to the largest corrections to the measured resonance frequencies, the absence of a viscous boundary layer and the small magnitude of the thermal boundary layer correction in the sphere also contribute to the attainment of high accuracy in the sound speed. Furthermore, small boundary layer losses mean that absorption in the bulk of the gas can be resolved from the linewidths of the acoustic resonances with greater precision than could otherwise be attained.

Insensitivity to geometric imperfection is important when absolute accuracy is required of the sound speed. Because the perturbation caused by some distortion described by a small parameter ε is only of order ε^2 [29], one can obtain speeds of sound accurate to parts per million using a resonator with geometric integrity at the level of parts in ten thousand. This is fortunate because it turns out that tolerances of around 10 μm in spheres with diameter of order 100 mm are attainable at modest cost using conventional machine tools but that substantially higher accuracy is likely to be difficult and/or expensive.

A further advantage of the spherical geometry lies in the availability of closed-form solutions to the problem of coupling between fluid and shell motion [30]. Such coupling becomes important as the gas density is increased and probably becomes the main factor controlling the accuracy attainable at pressures above a few megapascal.

6.4.2 Review of the acoustic model

In the theory of the spherical resonator, described in Chapter 3 and elsewhere [25], the resonance frequencies and half-widths can be written in the following form:

$$f_{ln} + \mathrm{i}g_{ln} = v_{ln}(u/2\pi a) + \sum_j (\Delta f + \mathrm{i}g)_j. \tag{6.4.1}$$

Here, v_{ln} is the eigenvalue of the (l, n) mode and $\sum_j (\Delta f + \mathrm{i}g)_j$ is the sum of small corrections that account for the approximations in the simplest treatment. The principal correction terms are those arising from the boundary layers, the coupling of fluid and shell motion, and from dissipation throughout the volume of the fluid. Other corrections may be included to account for holes, slots or cracks in the resonator's wall and for geometric imperfections.

As an illustration of the magnitude of the principal correction terms for the radial modes of the sphere resonator, figure 6.8(a) shows as a function of pressure the fractional thermal boundary layer and shell corrections to the frequencies of the $(0, 2)$ and $(0, 5)$ radial modes of a typical aluminium alloy sphere with internal diameter 100 mm and wall thickness 10 mm containing methane at 300 K. In figure 6.8(b) the contributions of boundary layer and bulk dissipation to the resonance half-widths are illustrated for the same fluid and resonator. Methane exhibits a very large bulk viscosity due to vibrational relaxation ($\eta_b/\eta \approx 500$ [31]) and for many gases the bulk absorption will be about two orders of magnitude smaller.

Often a small tubular hole is present in the wall of the resonator through which the fluid under study is admitted. This leads to a small perturbation to the resonance frequencies and half-widths given, for the radial modes, by equation (3.7.23); the specific acoustic input admittance y_0 of the opening which appears in that equation may be obtained from equation (5.8.20). In any event, it is easy to choose the length L and area ΔS of the inlet tube such that the shift in the resonance frequency does not exceed a few parts in 10^5 but, by a clever choice of length, the correction can be made even smaller for the radial modes [31]. In this design, which exploits the fact that the eigenvalues v_{0n} are approximately equal to $(n - \frac{1}{2})\pi$, the length is made equal to the radius of the sphere so that $\mathrm{Re}(\mathrm{i}y_0)$ vanishes near to the frequency of each radial mode. The exact length of the tube can be modified slightly to take account of end corrections and the departures of the eigenvalues from the approximate formula.

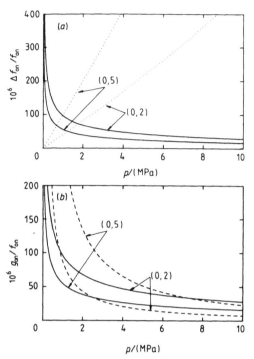

Figure 6.8 Fractional contribution to the (*a*) resonance frequencies and (*b*) half-widths of the $(0,2)$ and $(0,5)$ modes of a spherical resonator ($a = 50$ mm) containing methane at 300 K: $\cdots\cdots$, $-\Delta f_{\mathrm{sh}}$; ————, $-\Delta f_{\mathrm{h}} = g_{\mathrm{h}}$; —————, g_{b}.

Corrections for imperfect geometry can be applied if there is sufficient information to characterize the distortion. The theory has been worked out for the imperfections most likely to occur in practice, including spheroidal distortions and both misalignment and/or unequal radii of two hemispheres from which the resonator is constructed [29, 32]. Information about imperfect geometry can be obtained from mechanical measurements, from measurements of the splitting of the $(2l + 1)$ components of a nonradial acoustic mode, or from similar splitting observed in the microwave spectrum of the cavity [26, 33]. None of these methods is likely to provide complete information. In particular, certain geometric imperfections that cause second-order shifts in the frequencies of the radial modes do not cause corresponding first-order splitting of either the non-radial modes or the microwave modes [32]. Ideally, the theory should be used to establish the tolerances required for negligible shape perturbations and the resonator should then be manufactured to within those limits.

6.4.3 Excitation and detection of resonances

Although a single transducer could, in principle, suffice, practical systems have employed two—one to excite the resonances and one to detect them. Since, for the radial modes, the wavefunctions are uniform in values over the surface, any pair of locations on the wall would be equally efficient. This is not, of course, true for the non-radial modes and it is possible to exclude some of them altogether by placing the detector at a nodal angle with respect to the source. Thus, it has been found useful to separate the source and the detector by an angle of $\pi/2$ so that the $(0, 2)$ radial mode does not suffer interference from the nearby $(3, 1)$ non-radial mode which lies at a frequency only about 0.5 per cent higher. If both transducers are small then complete cancellation should be possible in a perfect sphere where there is no intrinsically preferred alignment for the 'polar' axis of the non-radial modes. In that case, the source transducer must establish the axis itself. However, if the sphere is imperfect then the perturbed non-radial modes may have a preferred axis in the cavity not necessarily coincident with the source. Together with the non-zero size of real transducers, this probably accounts for the incomplete cancellation of the $(3, 1)$ mode seen in practice.

In a typical approach to data collection [21], the excitation frequency is increased in discreet steps of $g_N/5$, from about $f_N - g_N$ to about $f_N + g_N$, and, at each point, the in-phase and quadrature components of the received signal are measured with a lock-in amplifier. As a counter-measure against any drifts in temperature or other properties during the data collection period, the resonance may be scanned first with increasing and then with decreasing frequency. The average of the two scans will then correspond to the conditions prevailing when the sign of the frequency step was reversed and should be independent of any drifts that are linear in time. In many cases, much simpler instrumentation, such as that described later in §6.5.1, could be used to measure the resonance frequencies with useful accuracy.

We have already seen that the acoustic pressure at the detector is essentially a Lorentzian function of frequency near to a resolved singlet. In practice, there is some background signal from the tails of adjacent resonances and (possibly) electrical and mechanical crosstalk between source and detector. Data collected near to a resonance may be analysed numerically using equation (3.3.10) in which the resonance is described by the two constants A_N and $F_N = f_N + ig_N$ [34]. The amplitude A_N of the resonance is taken as a complex constant because there are almost always uncontrolled phase shifts present in the source–detector path. When the background signal is only a few per cent of the resonance signal or less, it is usually sufficient to include just the constant background parameter B in the analysis so that a fit containing only three complex parameters is required. Inclusion of the leading frequency-dependent term $C(f - f_0)$ in the background has allowed background levels comparable in amplitude with the resonance term itself to be tolerated. Although best accomplished using both amplitude and phase information collected at a number of frequencies around resonance, analyses using amplitude–frequency data alone would suffice in many cases of low background signal and high resonance quality factor.

When the 'background' signal is attributable mainly to interference with adjacent modes, it is possible to subtract that part of the signal analytically using data collected over both resonance curves. In one method that has been used to counter overlap between the $(0, 2)$ radial mode and the $(3, 1)$ non-radial mode, the two sets of data are each analysed separately, as singlets with constant background, after subtraction of the estimated contribution of the other [35]. Several iterations are required to establish the optimum parameters for both resonances. This will lead to reliable results for the radial mode when the components of the adjacent non-radial mode, seven of them for the $(3, 1)$ mode, are split by an amount small compared with the half-widths (thus justifying their treatment as a singlet). When the overlapping modes are superficially unresolved or only partially resolved, the whole multiplet must be treated together in one non-linear regression analysis. The three components of some $(1, n)$ multiplets have been analysed successfully in this way for the case of a resonator deformed sufficiently to split them by an amount of order 100 PPM [25].

6.4.4 Non-radial modes

The fact that the individual components of the $(1, n)$ and possibly also the $(2, n)$, multiplets can be resolved in practice raises the issue of to what use the non-radial modes may be put. Since the unweighted average of the resonance frequencies of the components of a multiplet has the same insensitivity to geometric imperfections as do the radial modes, we can see that accurate sound speeds could be obtained. But, compared with a radial mode, the analysis involves many more parameters, each of which is burdened by a correspondingly larger uncertainty, and the quality factors will be lower on account of viscous damping at the surface. Thus, while the non-radial modes may provide useful information about the geometric integrity of the resonator, they do not offer any advantages over the radial modes for either accurate or rapid determination of the sound speed. In principle, it is possible to determine the viscosity of a gas from the half-widths for non-radial modes (after allowance for bulk losses and losses in the thermal boundary layer). The sphere is, however, a poor geometry for this purpose. Not only are the modes somewhat complicated to analyse but the contribution of viscous damping in the boundary layer can only be increased by using a smaller sphere and that has the effect of increasing the frequency and hence the bulk losses. With these points in mind, non-radial modes will be considered no further.

6.4.5   A design example

The most accurate measurements of sound speed yet made in a gas were performed using the spherical acoustic resonator of Moldover *et al* during their redetermination of the gas constant [27, 28]. The measurements were performed in argon gas at 273.16 K and pressures between 25 and 500 kPa. The apparatus, illustrated in figure 6.9, provides an example of what can be achieved when the primary objective is attainment of the highest possible absolute accuracy.

In order to maximize the Q and minimize the thermal boundary layer correction, a large resonator was chosen ($a = 89$ mm, $V = 3$ l). It was fabricated in two hemispheres from an austenitic stainless steel using a numerically controlled lathe. The outer surface, which was cylindrical in shape either side of the joint between the two halves, was machined accurately concentric with the inner surface in a single pass of the turning tool and used for alignment purposes when the two hemispheres were bolted together. The internal surface was polished to a mirror finish after turning so that surface irregularities were small compared with the thickness of the thermal boundary layer. A small cylindrical extension initially present on each hemisphere was then removed by grinding, leaving smooth flat mating surfaces. The largest known geometric imperfection in the hemispheres was a difference in radii of about 23 μm (approximately two parts in 10^4).

Stainless steel was chosen as the material of construction primarily for its compatibility with mercury, which was used to calibrate the volume of the assembled resonator. Since the resonator was to be calibrated in this way, it was essential that is be sealed at the equatorial join without any cracks or crevices into which mercury might flow. This was achieved by depositing a thin film (≈ 3 μm thick) of a vacuum wax on to one of the mating surfaces which was then melted after assembly of the resonator and observed to flow under the pressure of the bolts. With the high values of the characteristic acoustic impedance $\rho_w u_w$ of stainless steel and the thick wall (19 mm), the shell correction was also kept very small.

The transducers were commercially available capacitance microphones, with flat nickel diaphragms, mounted in specially fabricated brass housings flush with the internal surface of the resonator. The gas inlet port was of a comparatively large diameter but was provided with a valve which, when closed, plugged the port leaving a smooth internal surface.

Using this apparatus, the speeds of sound obtained from the lowest five radial modes at each pressure has a typical range of less than 1 PPM after correction according to the acoustic model, and the zero density limit of u^2 was established with an imprecision of 0.7 PPM (one standard deviation). By far the largest correction term in the sound speed was that for the thermal boundary layer which amounted to 63 PPM averaged over the $(0, 2)$–$(0, 6)$ modes at 100 kPa; this is much smaller than the corresponding correction (about 1600 PPM) required for the cylindrical interferometer of Quinn *et al* [8] operating under similar conditions. The accuracy of the measurements was determined by a number of factors, including the accuracy of the volume, and the final value of the gas constant was associated with a fractional standard deviation of 1.7 PPM including all known random and systematic uncertainties.

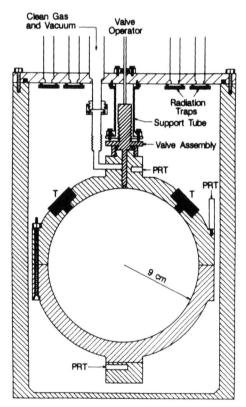

Figure 6.9 The spherical resonator of Moldover *et al* [28]: T, location of transducer assemblies; PRT, location of capsule thermometers.

It is also interesting to examine the amount Δg_{0n} by which the experimentally observed half-widths exceed those calculated from equation (3.7.15). The losses in the thermal boundary layer were dominant under the experimental conditions and so this comparison can be viewed as a test of the boundary layer theory. For the $(0,2)$–$(0,5)$ modes, $\Delta g_{0n}/f_{0n}$ increased slowly and roughly in proportion to $\sqrt{p}$ from around 1×10^{-6} at 25 kPa to about 3×10^{-6} at 500 kPa. The $(0,6)$ mode, however, showed evidence of coupling to a damped resonance of the shell; $\Delta g_{06}/f_{06}$ increased almost linearly with pressure reaching about 7×10^{-6} at 500 kPa. The small magnitude of the excess half-widths at low pressures was taken as evidence that the boundary layer theory is essentially correct but the residual discrepancies observed in that and other work have yet to be explained.

In other applications of spherical resonators, especially in the study of gas-phase equations of state [21, 31, 35], absolute accuracy is of less importance than the ability to operate with high precision over a wide range of temperature and pressure. Recent applications have employed spherical resonators at temperatures down to 100 K [36] and pressures up to 7 MPa [31]; further extensions of the operating conditions would seem likely. For applications in primary thermometry [37], ratios of sound speeds are required between different temperatures. Consequently, it is necessary to know the integrated thermal expansion of the resonator over the range in question and in this context measurements of the frequencies of the microwave resonances of the evacuated cavity look promising [33]. For more routine measurements, where both R and T are taken as known quantities, the mean radius of the resonator can be obtained from calibration measurements with a gas, such as argon, for which the ratio γ^{pg}/M, and hence the zero-density limit of the sound speed $(RT\gamma^{pg}/M)^{1/2}$, is known with high accuracy.

6.5 Cylindrical Resonators

Compared with spherical resonators, cylinders are easier to fabricate and offer an extra degree of freedom in the form of the ratio b/L which can be chosen to suit a particular application best. However, as already indicated, the resonances will not be as sharp in the cylindrical cavity as they would be in a comparable spherical resonator.

6.5.1 Excitation and detection of resonances

The location of the transducers used to excite and detect the resonances will depend upon which modes are of interest, with the most efficient locations at maxima of the corresponding wavefunction. For longitudinal modes, all locations on the end plate are equally efficient and a source located in the centre of an end plate will couple efficiently to all axisymmetric modes including pure radial and radial–longitudinal compound modes. A large plane source covering all or most of one end plate will excite longitudinal modes with great efficiency, often without significant interference from radial modes, and give a clean spectrum of resonances over a wide frequency range. Solid-dielectric capacitance transducers are particularly advantageous for this purpose and were incorporated in, for example, the apparatus of Younglove and McCarty [38] illustrated in figure 6.10. A possible disadvantage of such apparatus is that the impedance of the transducers, which may not be easy to calculate, becomes important in determining the resonance frequencies and half-widths. Excitation of non-axisymmetric modes in a cylindrical resonator requires a source offset from the centre of an end plate (because $J_m(0) = 0$ when $m > 0$) or else one located on the side wall. Each of those modes is twofold degenerate in a perfect cylinder.

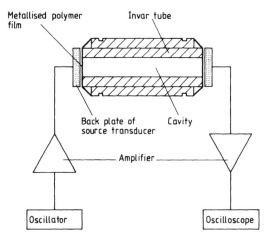

Figure 6.10 Cylindrical resonator and instrumentation as used by Younglove and McCarty [38].

Resonance data collection and analysis can be performed in exactly the manner described above in connection with the spherical resonator. Alternatively, resonance frequencies can be measured to within a few per cent of the linewidth using simple equipment: a variable-frequency oscillator, amplifier and frequency counter on the source side and an amplifier and oscilloscope on the detector side. One then adjusts the frequency of the source until a maximum in the received amplitude is observed on the oscilloscope; the resonance frequency may then be read from the counter. Combined with a mechanical measurement of the length and/or radius of the cylinder, or with a calibration measurement, this method will often be sufficient to obtain the sound speed with 0.1 to 0.01 per cent precision. Of course, these simple techniques could equally well be applied with any other resonator geometry.

6.5.2 Review of the acoustic model

The acoustic model by which the speed and absorption of sound is related to the resonance frequencies and half-widths in the cylindrical resonator is basically the same as for the spherical case. The important surface and bulk perturbations have been examined in Chapter 3 and, except for the quantitative estimation of the effects of coupling between fluid and shell motion, the theory is well developed. For the non-degenerate modes, it is easy to include the effects of coupling tubes and/or other openings in the wall.

6.5.3 Determination of transport coefficients of gases

Clearly, the cylindrical resonator is well suited to measurement of the speed of sound in fluids and both longitudinal and radial modes have been used for this purpose [38, 39]. The cylindrical geometry is also particularly suitable for determination of the thermal conductivity and viscosity of gases at low pressures. The existence of modes of differing symmetry allows the contributions of the two loss mechanisms to be separated. In polyatomic gases, the term $(\gamma - 1)$ can be quite small so that the viscosity would be determined more easily than the thermal conductivity. An advantage of the acoustic method over conventional ones is that molecular slip and the temperature jump do not affect the half-widths in leading order. Consequently, measurements should be possible at quite low pressures where the mean-free-path effects are a problem in the usual methods of determining the transport coefficients. A precision of about 0.1 per cent in the half-width can be obtained in practice without much difficulty so that random errors in the measurements would compare favourably with the conventional techniques. But, weighing against these advantages, there is the problem of unexplained residual loss mechanisms operating in the resonator. We have already seen with the spherical resonator that such unexplained losses exist even in the most carefully designed experiment. Since the surface damping is considerably greater in a typical cylinder, the residual losses may be comparatively smaller. The surface losses $(\sim f^{1/2}/\rho^{1/2})$ can be separated from those of bulk losses $(\sim f^2/\rho)$ either by a calculated correction for the latter or, when the bulk viscosity is unknown, by a regression analysis in which terms of the appropriate frequency and density dependence are adjusted in the fit.

As a numerical example, consider a cylindrical cavity of length 30 mm and radius 10 mm filled with butane gas at 300 K and 0.1 MPa. Under these conditions, the speed of sound is 211.5 m s^{-1}, $\gamma = 1.107$, and the viscous and thermal diffusivities are about 3.2×10^{-6} and 4.0×10^{-6} m^2 s^{-1} respectively [40]. The second longitudinal mode will lie at 7050 Hz with a theoretical half-width of 5.10 Hz, while the lowest radial mode will lie at

12 890 Hz with a theoretical half-width of 2.91 Hz. Since the ratio g_h/g_s turns out to be 0.20 for the longitudinal modes and 0.48 for the radial modes, and the bulk losses are very small in this frequency range (contributing only about 3 per cent of g for the radial mode), separation of the viscous and thermal diffusivities should not be difficult in this case. Notice, however, that to obtain this information the heat capacity ratio γ is required, and that to obtain x and η themselves it is necessary to know the density and specific heat capacity of the gas. Corrections for the non-ideality of the gas can be significant for the thermodynamic properties required here; knowledge of the second virial coefficient B and its first two temperature derivatives will be required for the most accurate work. It may be advantageous in practice to reduce the pressure further so that the interesting losses are enhanced relative to any residual mechanisms even though the bulk dissipation would increase more than the surface dissipation. In any event, measurements over a range of pressures, and with different modes, would be desirable in order to test the theoretical frequency and density dependence of the half-widths. Such measurements would also provide the thermodynamic information required in the form of the zero-density heat capacities and the second acoustic virial coefficient (from which B and its temperature derivatives may be obtained).

Small cylindrical resonators have been applied by Carey *et al* in the determination of sound speed and transport coefficients for nitrogen, argon, hydrogen, and air over a very wide temperature range (up to 1000 K) and at both high temperature and high pressure (up to 620 K and 50 MPa) [39]. Accuracies of 0.1 per cent in η and 0.5 per cent in x were claimed at low densities; these claims are borne out by comparison with 'conventional' measurements of comparable accuracy near 300 K [46].

6.6   Numerical Methods

With the availability at low cost of laboratory computers for data acquisition and analysis, numerical methods are a much more attractive means of handling resonance data than the traditional graphical ones.

6.6.1   Fitting functions

The numerical problem is almost the same for measurements of either impedance as a function of length or of acoustic pressure as a function of length or of frequency. For the variable path interferometer, the measured impedance or acoustic pressure near to resonance takes the form of the Lorentzian functions given by equation (6.2.7) or (6.3.2), with or without the addition of a constant background term. The complex signal $w(L)$, corresponding to Z_m or p_a as functions of length, therefore requires analysis in terms of the equation

$$w(L) = \left(\frac{A}{\Delta L_n^r + i(L - L_n^r)}\right) + B. \tag{6.6.1}$$

Generally, A and B must be taken as adjustable complex parameters, although if the experimenter is confident that no spurious phase shifts exist in the measurement system then the phase of A might be known (e.g. with $w \sim Z_m$, A should be purely real).

For variable-frequency resonance measurements near to a resolved singlet in a constant-volume cavity, the measured signal $w(f)$ requires analysis in terms of the function

$$w(f) = \left(\frac{A'}{(F_N/f)^2 - 1}\right) + B + C(f - f') \tag{6.6.2}$$

which, as discussed in Chapter 3, reduces to the Lorentzian function

$$w(f) \approx \left(\frac{A}{g_N + i(f - f_N)}\right) + B + C(f - f') \tag{6.6.3}$$

for a high Q resonance.[5] The difference between the Lorentzian functional form, which appears for the variable pathlength interferometer, and the form strictly applicable in the present case is attributable to the inclusion or exclusion of certain normal modes in the 'resonance' term. In the case of equation (6.6.3), all of the longitudinal modes are explicitly included, while in (6.6.2) only a single normal mode is included and all others are considered as background. Consequently, for a cylindrical resonator driven by a transducer that covers one entire end plate, the Lorentzian form might actually be preferable. However, since in leading order the resonance term of equation (6.6.3) differs in its frequency dependence from that of (6.6.2) only by the factor $1 + (f - f_N)/2f_N$, which will usually differ insignificantly from unity, either function should be adequate.

6.6.2 Linear fitting procedures

If the background signal is genuinely negligible and the Lorentzian lineshape applies then a very simple linear analysis is possible because

$$w(x)^{-1} = A^{-1}(\Delta x_0 - ix_0) + iA^{-1}x \tag{6.6.4}$$

where x is the independent variable (L or f), x_0 is the resonance length or frequency, and Δx_0 is the half-width. The χ^2 goodness-of-fit statistic may be generalized to the case of a complex dependent variable y with the definition

$$\sum_{k=1}^{M} [(y_k - y_{fit,k})(y_k^* - y_{fit,k}^*)]/(M - m) \tag{6.6.5}$$

in which y_k $(k = 1, 2, ..., M)$ are the experimentally determined values of y at the M values of the independent variable, $y_{\text{fit},k}$ denotes the corresponding values given by the fitting equation, and m is the number of parameters. Using this definition, extension of the usual linear regression methods to cope with a complex dependent variable such as w^{-1} is straightforward. Consequently, it is a simple matter to obtain the two complex parameters A^{-1} and $(\Delta x_0 - i x_0)$ in equation (6.6.4).

When the background signal is small but not negligible, $w(x)^{-1}$ may be linearized, correct to leading order in the background, to obtain either a quadratic $(C = 0)$ or a cubic $(C \neq 0)$ polynomial in x, the coefficients of which give x_0, Δx_0 and the other parameters. This method can also be extended into an iterative algorithm capable of handling large background levels without approximation [41].

Often only the magnitude of w will be measured and, when the background signal is small and constant, a linearization of $|w|^{-2}$ correct to first order in B is appropriate [34]. This also gives a polynomial in the frequency

$$|w|^{-2} = c_0 + c_1 f + c_2 f^2 + c_3 f^3 \qquad (6.6.6)$$

in which

$$
\begin{aligned}
c_0 &= |A|^{-2}[x_0^2 + (\Delta x_0)^2]\sigma \\
c_1 &= |A|^{-2}\{-2x_0\sigma + [x_0^2 + (\Delta x_0)^2](2b_2/a)\} \\
c_2 &= |A|^{-2}[\sigma - 4(b_2 x_0/a)] \\
c_3 &= |A|^{-2}(2b_2/a) \\
\sigma &= 1 - 2(b_1\Delta x_0/a) - 2(b_2 x_0/a) \\
B &= b_1 + i b_2.
\end{aligned}
\qquad (6.6.7)
$$

Consequently, standard linear regression methods can be used to obtain precise values of the resonance length or frequency and the half-width from measurements of $|w(x)|$ in the presence of background levels up to a few per cent of the resonance amplitude $|A|/\Delta x_0$. It is interesting to note that the maximum amplitude of the experimental quantity occurs, in the presence of background, not at exactly x_0 but at

$$x = x_0 - b_2[(\Delta x_0)^2/|A|]. \qquad (6.6.8)$$

Consequently, identification of the maximum amplitude with resonance would lead to an error in the resonance length or frequency equal to the product of the fractional out-of-phase background signal $b_2(\Delta x_0/|A|)$ and the resonance half-width [34].

6.6.3   Non-linear analysis

When the background signal is more than a few per cent of the total, linearization of the fitting function will be inaccurate and a full non-linear regression analysis should be adopted. This can be performed using just the amplitude $|w(x)|$ but it is much better to measure the in-phase and quadrature components of the signal and use both of them in the analysis. Typically, the real and imaginary components of $w(x)$ are measured for a number of discrete values of the argument x within a half-width either side of the resonance. A precision of between 0.1 per cent and 0.01 per cent of the maximum amplitude can be achieved in practice using a lock-in amplifier. The most versatile method of analysis is the non-linear regression algorithm developed by Mehl and described in [34]. Essentially, this is the generalization to a complex-dependent variable of the widely used methods discussed by Brevington [42]. Initial estimates of the parameters are obtained by, for example, the linear fitting method described above. The function $w(x)$ is then linearized about the starting point in parameter space and an improved set of parameters found by solving a matrix equation. This procedure is iterated until the parameter set corresponding to the minimum χ^2 is found. Usually, with reasonable initial estimates, the method converges quickly to the correct solution. Occasionally with difficult cases (e.g. with a strong frequency-dependent background) the calculated increments may lead to an increase in the value of χ^2 rather than a reduction. In that case, the increments can be divided by some arbitrary factor until a set that reduces χ^2 is obtained before starting the next iteration (alternatively, the Levenberg–Marquardt algorithm can be employed [42]). In difficult cases, the number of iterations required for convergence can be dramatically reduced with good estimates of the starting parameters. One useful method exploits the fact that, unlike $|w|$ itself, $|dw/dx|$ has a maximum at resonance which is independent of any constant background level [41]. In terms of the Lorentzian functional form, this derivative is given by

$$|dw/dx| = |A|/[(\Delta x_0)^2 + (x - x_0)^2] \qquad (6.6.9)$$

and in practice both x_0 and Δx_0 can be estimated quite precisely from a table of first-order finite differences. Having estimated these two quantities, the parameters A and B may be obtained by linear regression with the variable $y = [\Delta x_0 + i(x - x_0)]\,w$:

$$y = [A + B(\Delta x_0 - ix_0)] + iBx. \qquad (6.6.10)$$

In this way, the full non-linear analysis can be initiated with greatly improved starting estimates for the parameters.

In variable-frequency experiments, the frequency-dependent background term $C(f - f')$ may be significant and it is best to fit the data first with C constrained to zero and then again with that parameter included in the fit. Standard statistical tests can be used to establish whether any resulting reduction in χ^2 is significant. The more complicated background signals that arise in the presence of significant overlap between adjacent modes are best handled using multiple resonance terms.

With a precision of 0.1 per cent in the measurements of the real and imaginary components of w, the interesting resonance parameters x_0 and Δx_0 can be obtained with a precision of about 0.2 per cent of Δx_0.

6.7 Choice of Methods

In the discussion of experimental methods, an attempt has been made to do justice to the capabilities of all the principal constant wave techniques, both old and new. The question of course arises as to which of them is best for a given application and a few comments in answer are perhaps appropriate.

There is little doubt that, for gases under conditions of low or moderate pressures not too close to saturation, the highest possible precision in the speed of sound will be achieved using a spherical resonator. Not only are the boundary layer losses small but the frequencies are low too, allowing measurements to be performed on many polyatomic gases without incurring problems with dispersion. The small surface damping also lends the method to low-frequency measurement of bulk viscosity in gases. As we have seen, the method can also yield very high absolute accuracy in the sound speed.

If the requirement is for simple equipment offering more moderate accuracy then a variable-frequency cylindrical resonator, combined with the modest instrumentation described in §6.5.1, would be a good choice.

In the study of sound speed and absorption at high frequencies, the double-transducer interferometer, being useful at both high and low standing wave ratio, probably eclipses the conventional single-transducer design. Although not immune, the interferometer may be less sensitive to systematic errors caused by precondensation on the walls than would a resonator operating at low frequencies. Since the method also functions well in the liquid phase it is a useful technique for studies ranging across the phase boundaries. With careful attention to the minimization of diffraction errors, an unguided interferometer operating at frequencies of a few megahertz could give excellent results in liquids or gases essentially free from both viscothermal and precondensation effects.

For measurements of the transport coefficients of gases, the various modes of a low-frequency cylindrical resonator offer a favourable selection of viscous and thermal surface damping terms combined with low bulk absorption. The method has not been very widely used; nevertheless, it would appear to have advantages over conventional techniques that make it worthy of further investigation.

References

[1] Hubbard J C 1931 *Phys. Rev.* **38** 1011
[2] Hubbard J C 1932 *Phys. Rev.* **41** 523
[3] Colclough A R 1973 *Metrologia* **9** 75
[4] Fay R D and White J E 1948 *J. Acoust. Soc. Am.* **20** 98
[5] Smith D H and Harlow R G 1963 *Br. J. Appl. Phys.* **14** 102
[6] Harlow R G and Kitching R 1964 *J. Acoust. Soc. Am.* **36** 1100
[7] Colclough A R 1979 *Acustica* **42** 18
[8] Quinn T J, Colclough A R and Chandler T R D 1976 *Phil. Trans. R. Soc.* A **283** 367
[9] Zartman I F 1959 *J. Acoust. Soc. Am.* **21** 171
[10] Parbrook H D 1953 *Acustica* **3** 49
[11] Blitz J 1967 *Fundamentals of Ultrasonics* 2nd edn (London: Butterworths) pp 117–18
[12] Bechmann R 1952 *J. Sci. Instrum.* **29** 73
[13] Colclough A R 1979 *Proc. R. Soc.* A **365** 349
[14] Colclough A R, Quinn T J and Chandler T R D 1979 *Proc. R. Soc.* A **368** 125
[15] Colclough A R 1979 *Acustica* **42** 28
[16] Henderson M C and Peselnick L 1957 *J. Acoust. Soc. Am.* **29** 1074
[17] Gammon B E and Douslin D R 1970 *Proc. Fifth Symp. on Thermophysical Properties* ed. C F Bonilla (New York: The American Society of Mechanical Engineers) pp 107–14
[18] Gammon B E 1976 *J. Chem. Phys.* **64** 2256
[19] Gammon B E and Douslin D R 1976 *J. Chem. Phys.* **64** 203
[20] Moldover M R, Waxman M and Greenspan M 1979 *High Temp. High Press.* **11** 75
[21] Mehl J B and Moldover M R 1981 *J. Chem. Phys.* **74** 4062
[22] Mehl J B and Moldover M R 1982 *Proc. Eighth Symp. on Thermophysical Properties, Volume 1: Thermophysical Properties of Fluids* ed. J V Sengers (New York: The American Society of Mechanical Engineers) pp 134–41.
[23] Mehl J B and Moldover M R 1982 *J. Chem. Phys.* **77** 455
[24] Moldover M R and Mehl J B 1984 *Precision Measurements and Fundamental Constants II* ed. B N Taylor and W D Phillips (Washington, DC: National Bureau of Standards) Special Publication 617, pp 281–6
[25] Moldover M R, Mehl J B and Greenspan M 1986 *J. Acoust. Soc. Am.* **79** 253
[26] Mehl J B and Moldover M R 1986 *Phys. Rev.* A **34** 3341
[27] Moldover M R, Trusler J P M, Edwards T J, Mehl J B and Davis R S 1988 *Phys. Rev. Lett.* **60** 249
[28] Moldover M R, Trusler J P M, Edwards T J, Mehl J B and Davis R S 1988 *J. Res. NBS* **93** 85
[29] Mehl J B 1982 *J. Acoust. Soc. Am.* **71** 1109
[30] Mehl J B 1985 *J. Acoust. Soc. Am.* **78** 782
[31] Goodwin A R H 1988 *PhD Thesis* University of London
[32] Mehl J B 1986 *J. Acoust. Soc. Am.* **79** 278
[33] Ewing M B, Mehl J B, Moldover M R and Trusler J P M 1988 *Metrologia* **25** 211
[34] Mehl J B 1978 *J. Acoust. Soc. Am.* **64** 1523

[35] Ewing M B and Trusler J P M 1989 *J. Chem. Phys.* **90** 1106

[36] Ewing M B, Owusu A A and Trusler J P M 1989 *Physica* A **156** 899

[37] Moldover M R and Trusler J P M 1988 *Metrologia* **25** 165

[38] Younglove B A and McCarty R D 1980 *J. Chem. Thermodyn.* **12** 1121

[39] Carey C, Bradshaw J, Lin E and Carnevale E H 1974 *Experimental Determination of Gas Properties at High Temperatures and/or High Pressures* (Arnold Engineering Development Center, Arnold Air Force Station, TN 37389, USA) *Report no.* AEDC-TR-74-33, also available as NTIS AD-779772

[40] Ewing M B, Goodwin A R H, McGlashan M L and Trusler J P M 1988 *J. Chem. Thermodyn.* **20** 243

[41] Ewing M B and Trusler J P M 1989 *J. Acoust. Soc. Am.* **85** 1780

[42] Brevington P R 1969 *Data Reduction and Analysis for the Physical Sciences* (New York: McGraw-Hill) pp 232–46

[43] Krasnushkin P E 1968 *Acustica* **20** 343

[44] Del Grosso V A 1966 *Systematic Errors in Ultrasonic Propagation Parameter Measurements Part 3—Sound Speed by Iterative Reflection Interferometry* (Washington, DC: Naval Research Laboratories) *NRL Report* 6409, also available as NTIS AD-647356

[45] Del Grosso V A 1971 *Acustica* **24** 299

[46] Maitland G C, Rigby M, Smith E B and Wakeham W A 1981 *Intermolecular Forces—Their Origin and Determination* (Oxford: Clarendon) p 571

[47] Del Grosso V A and Mader C W 1972 *J. Acoust. Soc. Am.* **52** 961

[48] Del Grosso V A and Mader C W 1972 *J. Acoust. Soc. Am.* **52** 1442

[49] Kroebel W and Mahrt K-H 1976 *Acustica* **35** 154

Notes

1. The definitions of resonance and anti-resonance are used here in the opposite sense to their usual meaning in circuit theory: both refer to one of the two points of zero phase on each impedance circle but resonance is taken as the one with the greater input resistance.
2. End-face corrections will be given in the following section.
3. Guided-mode dispersion is a term used to describe the effects of transmission modes other than the plane-wave mode on the apparent sound speed observed in a waveguide.
4. In that method, self-sustained oscillations were obtained using a feedback loop in which the amplified signal from the detector transducer was used to energize the source. Under conditions of steady-state oscillations, the total phase shift around the loop must be an integer multiple of 2π. Consequently, provided that there are no phase shifts in the electronics, the frequency of oscillation is mu/L, where m is an integer and L is the pathlength. m can be determined from the displacement necessary to reach the next separation at which the same oscillation frequency is obtained.
5. The factor i present in equation (3.3.11) has been absorbed into the constant A.

7

Experimental Methods 2: Transient Techniques

7.1 Introduction

Broadly, the transient methods to be described in this final chapter fall into two categories: pulse methods and reverberation techniques. In the former, the speed of sound is determined, often with very high precision, essentially from a measurement of a pulse propagation time over a known distance in the fluid. It is also possible to obtain the absorption coefficient from the attenuation of the pulse. Typically, pulse methods are used with carrier frequencies from about 1 MHz upwards. They function particularly well in dense fluids but can also be used in gases. The reverberation method is used to measure absorption of sound. In this method, one measures the logarithmic decrement for a freely oscillating normal mode of a resonator containing the fluid under study. Under favourable conditions this quantity is dominated by the absorption throughout the volume of the fluid and leads to a measurement of the bulk absorption coefficient. Reverberation techniques can be used at frequencies from the infrasonic to the ultrasonic, thus allowing relaxation mechanisms to be studied over many orders of magnitude in the time domain, and in either gases or liquids.

7.2 Pulse Methods

Many of the pulse techniques used for sound speed and absorption measurements in fluids have been adapted from the methods, which are fully described in the literature [1–3], for measuring the elastic moduli of solids. But experiments on fluids differ from those on solids in a number of respects. For example, unlike a fluid, a solid specimen can have anisotropic elastic properties. Another important distinction lies in the boundary conditions. Fluid samples are usually contained within nearly rigid boundaries, while pressure-release boundary conditions apply at the free surface of a solid specimen. Accordingly, the main pulse techniques will be examined here from the perspective of the experimenter interested in the speed and absorption of sound in fluids.

7.2.1 Pulse propagation

We begin with a few elementary points about the shape, propagation and reflection of sound pulses in fluids. An indication of the orders of magnitude involved is afforded by the delay time across a 1 cm path in a fluid with $u = 1000 \text{ m s}^{-1}$, which is 1 μs. Quite inexpensive interval timers can offer single-measurement resolution of order 1 ns while the more sophisticated ones can resolve events on a picosecond timescale. In practice one can usually obtain delay times by comparison (e.g. on an oscilloscope display) with the accurately known period of a clock signal. The measurement then becomes one of frequency. Consequently, very high resolution is available but the accuracy is controlled by detailed considerations relating to pulse shape distortion and to phase shifts associated with diffraction and reflections.

An ideal pulse of duration τ clipped from a continuous sinusoidal signal of angular frequency ω_0 is illustrated in figure 7.1(a). This may be represented by the imaginary part of the function

$$P(t) = w(t)\exp(\mathrm{i}\omega_0 t) \qquad (7.2.1)$$

in which $w(t)$ is a gating function defined by

$$\begin{aligned} w(t) &= 0 & t &< -\tfrac{1}{2}\tau \\ &= 1 & -\tfrac{1}{2}\tau &< t < \tfrac{1}{2}\tau \\ &= 0 & t &> \tfrac{1}{2}\tau. \end{aligned} \qquad (7.2.2)$$

The real part of $P(t)$ then represents a pulse clipped from a constant wave signal $\cos(\omega_0 t)$. Of course, several features of real systems prevent us from ever attaining the perfectly rectangular envelope of these ideals. For example, electronic gate circuits have finite rise and fall times. Actually, we can assume for most practical purposes that electronic circuits are perfect and concentrate instead on the other aspects of the experiment. To analyse the behaviour of the system in response to pulse excitation we should consider the Fourier transform $S(\omega)$ of the function $P(t)$:

$$S(\omega) = \frac{1}{2\pi} \int_{-\infty}^{\infty} P(t)\exp(-\mathrm{i}\omega t)\, \mathrm{d}t = \frac{\sin\left[(\omega - \omega_0)\tau/2\right]}{\pi(\omega - \omega_0)}. \qquad (7.2.3)$$

In terms of this function, the power transmitted in the band between ω and $\omega + \mathrm{d}\omega$ is proportional to the product $|S(\omega)|^2\, \mathrm{d}\omega$, where $|S(\omega)|^2$ is the spectrum density of the pulse. The spectrum density, shown as a function of $(\omega - \omega_0)\tau$ in figure 7.1(*b*), is symmetric about $\omega = \omega_0$ where it is maximum. Since the half-power bandwidth is of order τ^{-1}, very short pulses contain frequency components over a broad band, while the frequency spectrum of a long pulse approaches the shape of the delta function $\delta(\omega - \omega_0)$.

Piezoelectric expander plates, usually of *x*-cut quartz, are often used as transducers. Typically, the carrier frequency is set equal to the resonance frequency of the transducer but, because of limited bandwidth, the envelope of the pulse transmitted into the fluid is significantly distorted; there is a finite rise time and ringing at the trailing edge. The detailed form of the transmitted pulse can be obtained from our knowledge of the behaviour under constant wave excitation by use of the inverse Fourier transform.

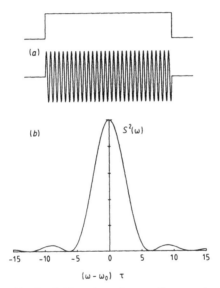

Figure 7.1 An idealized 30-cycle pulse cut from a sinusoidal signal: (*a*) time-domain representation $P(t)$; (*b*) spectrum density $|S(\omega)|^2$.

When a constant wave signal is applied to the transducer, the resulting displacement ξ is proportional to the current passing through the mechanical branch of the equivalence circuit (see figure 5.11) and, with an excitation voltage V at an angular frequency ω, this is given by

$$I = \left(\frac{V}{R_1}\right)\left(\frac{\omega\omega_0}{\omega\omega_0 + iQ(\omega^2 - \omega_0^2)}\right). \qquad (7.2.4)$$

The time-domain response $\xi(t)$ of the transducer under pulse excitation is therefore proportional to the inverse Fourier transform of $\Im(\omega)S(\omega)$, which we shall denote by $P'(t)$. Here, $\mathfrak{I}(\omega)$ is a dimensionless transfer function proportional to the admittance of the transducer and given by

$$\mathfrak{I}(\omega) = \frac{\omega\omega_0 Q}{\omega\omega_0 + iQ(\omega^2 - \omega_0^2)}. \tag{7.2.5}$$

When $P'(t)$ is expressed as the difference

$$
\begin{aligned}
P'(t) &= \int_{-\infty}^{\infty} \mathfrak{I}(\omega)S(\omega)\exp(i\omega t)d\omega \\
&= \int_{-\infty}^{\infty} \mathfrak{I}(\omega)\left(\frac{\exp\left[i\omega(t + \frac{1}{2}\tau) - i\frac{1}{2}\omega_0\tau\right]}{2\pi i(\omega - \omega_0)}\right)d\omega \\
&\quad - \int_{-\infty}^{\infty} \mathfrak{I}(\omega)\left(\frac{\exp\left[i\omega(t - \frac{1}{2}\tau) + i\frac{1}{2}\omega_0\tau\right]}{2\pi i(\omega - \omega_0)}\right)d\omega
\end{aligned} \tag{7.2.6}
$$

the transform can be evaluated exactly by contour integration. Each of the two terms into which $P'(t)$ is resolved has a simple pole at $\omega = \omega_0$ with residue $(Q/2\pi i)\exp(i\omega_0 t)$. In addition, each has simple poles at $\omega = (i\omega_0/2Q) \pm \omega_0'$; these are the poles of the transfer function and when $Q > \frac{1}{2}$, ω_0' is a real quantity given by

$$\omega_0' = \omega_0[1 - (1/4Q^2)]^{1/2} \tag{7.2.7}$$

and the residues are

$$\left(\frac{(i\omega_0/2Q) \pm \omega_0'}{\pm 2i(\omega_0'/\omega_0)}\right) \times \left(\frac{\exp[-(\omega_0/2Q)(t + \sigma\tau)]}{(i\omega_0/2Q) \pm \omega_0' - \omega_0}\right)$$

$$\times \left(\frac{\exp[\pm i\omega_0'(t + \sigma\tau) - i\omega_0\sigma\tau]}{2\pi i}\right)$$

where $\sigma = +\frac{1}{2}$ for the first integral and $\sigma = -\frac{1}{2}$ for the second. Since the integrands each have a pole on the real axis, the path of integration between $\omega = -\infty$ and $\omega = +\infty$ is made to take an infinitesimal detour passing just below $\omega = \omega_0$. The contour is then closed on a semi-circle of infinite radius extending from $\omega = +\infty$ back to $\omega = -\infty$ in either the upper or the lower half of the complex plane. When the path is closed in the upper half of the complex plane, all three poles are enclosed and the contour integral evaluates to $2\pi i$ times the sum of the residues, but when the path is closed in the lower half of the plane, all the poles are excluded and the integral is zero. The choice is dictated by the requirement that the integrand vanish along the semi-circle so that the contour integral is identical with the line integral in which we are interested. With that criterion, the first integral in equation (7.2.6) is zero for $t < -\frac{1}{2}\tau$ and non-zero for $t > -\frac{1}{2}\tau$, while the second integral is zero for $t < \frac{1}{2}\tau$ and non-zero thereafter. The resulting pulse function $P'(t)$ is, like $P(t)$, a complex quantity corresponding to an input pulse gated from a signal proportional to $\exp(i\omega_0 t)$. However, because of the symmetry properties of the transfer function $\mathfrak{I}(\omega)$, and the linear nature of the Fourier transform pair, the real and imaginary parts of $P'(t)$ can be identified with the responses to input pulses gated from constant wave signals $\cos(\omega_0 t)$ and $\sin(\omega_0 t)$ respectively.[1] Figure 7.2 shows pulses transmitted by transducers, having various values of Q, driven by perfect sinusoidal pulse trains 30 cycles long. When the normalization factor $(1/Q)$ applied in figure 7.2 is taken into account, the trade-off between pulse shape and amplitude becomes clear. For small values of Q, the bandwidth of the transducer is wide enough to transmit almost all components in the frequency spectrum of the excitation signal; the output pulse then rises rapidly to reach an essentially constant amplitude for most of its duration. On the other hand, a large value of Q can lead to a greatly increased maximum amplitude but, since the bandwidth of the device is narrow, only the centre of the frequency spectrum receives significant weight; the 'wings' on either side are attenuated relative to the carrier frequency and the pulse shape is badly distorted.

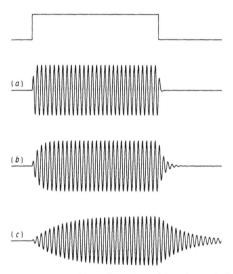

Figure 7.2 Distortion of a 30-cycle pulse arising from finite transducer bandwidth: (*a*) $Q = 0.8$; (*b*) $Q = 4$; (*c*) $Q = 20$. The pulse timing envelope is also shown.

Although this 'exact' evaluation of the pulse shape is simple enough, an approximation is satisfactory when $Q \gg 1$. This may be obtained by neglecting terms of order Q^{-1} and smaller so that $\omega_0' = \omega_0$ and the sum of residues simplifies to yield

$$
\begin{aligned}
P'(t) &= 0 && t < -\tfrac{1}{2}\tau \\
&= Q\{1 - \exp[-(\omega_0/2Q)(t + \tfrac{1}{2}\tau)]\}\exp(i\omega_0 t) && -\tfrac{1}{2}\tau < t < \tfrac{1}{2}\tau \quad (7.2.8) \\
&= 2Q\,\sinh(\omega_0\tau/4Q)\exp(-\omega_0 t/2Q)\exp(i\omega_0 t) && t > \tfrac{1}{2}\tau.
\end{aligned}
$$

The leading and trailing edges of the transmitted pulse are then characterized by the rise (and fall) time $\tau_r = 2Q/\omega_0$ while the maximum amplitude, reached at $t = \frac{1}{2}\tau$, is proportional to $Q[1 - \exp(-\tau/\tau_r)]$.

As an example, a thickness-mode quartz transducer resonant at the carrier frequency and operating in water ($Q \approx 8$) has a rise time of about $0.22 \, \mu s$ at 10 MHz ($\approx$ two cycles). It is advantageous, where possible, to select a transducer with its fundamental resonance, rather than a harmonic, at the desired frequency because this will offer greater bandwidth. The damping can be substantially increased by bonding a suitable material (tungsten-loaded epoxy resin, for example) to the rear face of the transducer. This is essential in the study of gases by pulse methods using resonant transducers although, in this case, solid-dielectric capacitance transducers offer an attractive alternative below 1 MHz [4, 5].

The pulse generated by the transducer can be subject to further shape distortion as it passes through the medium. A frequency-dependent absorption coefficient will attenuate some components of the frequency spectrum more than others while, in a dispersive medium, the different components will propagate with different phase speeds. To see if these are important effects we need to know whether α and/or u suffer significant variation over the bandwidth τ^{-1} centred on the carrier. Actually we can identify three distinct sound speeds in this context. The first is the phase speed, denoted by the undecorated u and equal to the speed at which points of equal phase propagate in a monofrequency wave. Second there is the group speed u_g, equal to the speed at which the centre of the pulse propagates. Third there is the signal speed u_s which is the speed at which the leading edge of the pulse propagates. When the absorption is small and the Fourier components of the signal have frequencies that do not differ too much from that of the carrier, the group speed is $u + \omega_0(\partial u/\partial \omega)_{\omega_0}$ [30]. One can also show that the signal speed is identical with the infinite frequency limit of the phase speed: $u_s = u(\omega \to \infty)$ [6]. Thus, in principle, one might expect to observe an apparent sound speed anywhere in a range from the group speed to the signal speed depending upon the threshold of the detector circuit. However, looking at the frequency spectrum of the pulse, one can see that the power transmitted at frequencies much higher than the carrier frequency is vanishingly small so that a real detector could not resolve the true leading edge of the received pulse. In practice, cycle-for-cycle matching between echoes, using the information in the centre of each pulse, leads to a good measure of the phase speed at the carrier frequency. On the other hand, significant errors are predicted in absorption coefficients of highly viscous liquids when narrow pulses are used [30].

We will also be interested in reflection from a solid surface. This too can introduce pulse shape distortion when the reflection coefficient $\chi_r = \exp(i\delta_r - \beta_r)$ exhibits significant variation over the pulse bandwidth. The reflected pulse can be obtained by taking the inverse Fourier transform of $\chi_r(\omega)S(\omega)$. When χ_r is frequency independent, the reflected and incident pulses have the same relation in amplitude and phase as do continuous waves, with attenuation $\exp(-\beta_r)$ and phase shift δ_r. When the reflector has a resonance near to the carrier frequency, the reflection coefficient can be strongly frequency dependent and lead to further distortion of the leading and trailing edges of the pulse; this is an important consideration especially when multiple echoes are exploited. The reflection coefficient for a plane reflector of thickness $2l$ can be estimated by combining equations (2.4.13) and (5.4.21); for a non-piezoelectric material ($K_t = 0$) at frequencies remote from a resonance that gives

$$\chi_r \approx 1 - i(Z_1/Z_2)[\tan(k_2 l) - \cot(k_2 l)] \tag{7.2.9}$$

where Z denotes characteristic acoustical impedance, k denotes propagation constant, and the subscripts 1 and 2 refer respectively to fluid and solid. By choosing a high-impedance material, such as tungsten, for the reflector and avoiding resonances χ_r can be made close to unity over the pulse bandwidth. Table 7.1 gives sound speeds and characteristic impedances for several possible reflector materials [7, 8].

Table 7.1 Properties of reflector materials.

Material	$u/(\mathrm{m\,s}^{-1})$	$Z/(\mathrm{MPa\,s\,m}^{-1})$
Brass	4372	40
Copper	4759	43
Stainless steel	5980	47
Tungsten	5221	100
Tungsten carbide	6655	104

7.2.2 Single-pulse methods

In the most elementary method, a delay line is used consisting of two plane parallel transducers placed a known distance apart in the fluid. Single pulses are applied to one transducer and detected by the other. The delay time can be measured in a number of ways including the use of an interval timer triggered by the transmitted and received pulses, and comparison on an oscilloscope of the received pulse with a second pulse delayed by an accurately known time [9, 10]. Many readings can be averaged by, for example, generating a continuous sequence of pulses on a clock cycle chosen so that all the echoes of the first have decayed before the second is initiated. Although high precision can be achieved, the measured time of flight will not be very accurate because of pulse shape distortion and non-zero delay times in the electronics. However, the method does allow changes in the sound speed to be measured with high precision [9, 10]. By operating with a variable pathlength [4, 5, 11], absolute measurements of the sound speed can be obtained with improved accuracy. The variable path method with pulse height measurement also provides precise results (perhaps good to 1 or 2 per cent) for the absorption coefficient free from errors due to reflection losses. Diffraction corrections, especially to the absorption coefficient, can be significant in unguided delay lines and will be considered in a later section.

Higher absolute accuracy in the sound speed can also be achieved by using multiple reflections of a single pulse travelling between a transducer and a plane parallel reflector. The transducer is switched from source to detection circuits after the pulse is transmitted so that the returning echoes can be observed. Using this method, the time between arrival of echoes of a single initial pulse can be obtained essentially free from errors due to delay times in the detection circuits.

By initiating source pulses on a regular clock cycle, simple equipment can be used to measure the echo delay time [3, 31]. The pulse repetition rate is made sufficiently low that all the echoes from one burst die out before the next one is initiated. It is convenient to gate the received signal so that only a pair of echoes are allowed to pass. The gated received signal is then displayed on an oscilloscope. In order to measure the delay time, a variable-frequency oscillator is used to trigger the horizontal axis of the oscilloscope and its frequency is adjusted until it is equal to an integer multiple of the inverse delay time between returning echoes. When that condition is exactly fulfilled, cycle-for-cycle overlap between each pair of echoes will be observed on the display. The delay time is then obtained by measuring the frequency of the time base. Synchronization of the pulse repetition, receiver gate, and oscilloscope time base to the same clock signal will ensure a steady display from successive bursts of the pulse generator. A precision of 0.01 per cent and an accuracy of about 0.1 per cent can be achieved in liquids. This echo overlap method has been exploited to advantage by Benson and his collaborators using a commercially available transducer and pulse generation/detection system [31].

As noted in Chapter 5, piezoelectric plate transducers, contacting the same medium on both faces, are inefficient reflectors of sound at their resonance frequencies. Consequently, transducers backed by the fluid under study are not very useful and the apparatus shown in figure 7.3 has found favour for measurements on liquids [12, 13]. Here, a quartz transducer is bonded to one end of a cylindrical metal buffer rod using a thin film of involatile fluid. The fluid under study fills the space between the other end of the rod and a plane parallel reflector. Sound pulses generated by the transducer are transmitted into the rod, from the rod to the fluid, and from the fluid to the reflector, but, at each interface, some of the sound energy is reflected back as echoes. The progress of a pulse is illustrated in figure 7.4. The returning echoes are detected by the same transducer, now switched to a detection circuit, and delay times are measured. The initial sound pulse transmitted into the buffer rod will probably have an unknown phase relation with the electronic excitation pulse and a significant rise time. But, with a bit of luck, reflections at the fluid–solid interfaces will involve only very small phase changes and little further distortion. Consequently, the time delay between the return to the transducer of, say, the first and second echoes from the reflector will be a good measure of the round trip transit time across the fluid path. The finite rise time of the transducer, and the propagation delays in the buffer rod and in the bond between rod and transducer, cancel in this measurement. The echo overlap method can be used with the received signal gated so that only echoes returning from the fluid–reflector interface reach the oscilloscope display.

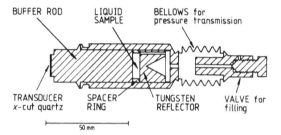

Figure 7.3 Apparatus for pulse–echo measurements on fluid samples at high pressures [13].

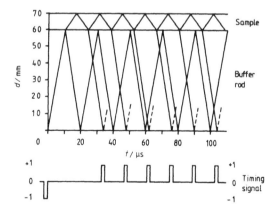

Figure 7.4 Time–distance plot for pulse–echo experiment. A gate timing signal $w(t)$ is also shown with $w = -1$ during pulse generation and $w = +1$ during pulse detection; when $w = 0$ both generator and detector gates are closed.

In the apparatus described by Petitet *et al*, and used at pressures up to 350 MPa, a 50 mm long buffer rod was used to couple a 10 MHz quartz transducer to a sample space just 5 mm long; the reflector was of tungsten [13].

7.2.3 *Double-pulse methods*

A number of techniques offering higher resolution and higher absolute accuracy have been developed for use on solid specimens and adapted for studies of fluids. In the phase cancellation method, which we describe first for use on a solid, a delay line is constructed by bonding transducers to opposite parallel faces of the sample [1]. Pairs of pulses with fixed width but variable time-domain separation are gated from a constant wave signal source and applied to the source transducer. Each of the pulses gives rise to multiple echoes travelling back and forth in the sample and the delay between the first and second pulse is adjusted until it is approximately equal to the round trip delay time in the sample. When that condition is achieved, the echoes of both pulses overlap and, being phase coherent, interfere with each other. With a little fine tuning of the carrier frequency, and some attenuation of the second pulse to allow for the extra attenuation experienced by the echoes of the first pulse, the echoes can be made to cancel each other out. This condition can be achieved, for example, by monitoring the received signal on a fast oscilloscope triggered by the gate signal.

It is worth examining this method a little more closely. If their width is sufficient then flat-topped pulses will be obtained and the constant wave analysis of Williams and Lamb can be adopted [1]. Ignoring the attenuation for the moment, we let the second and first pulses be $A \sin(\omega t)$ and $A \sin[\omega(t - t_0) + 2\phi_r]$ respectively, where t_0 is the round trip transit time and ϕ_r is the phase shift on reflection. The sum of these two signals is

$$2A \sin[\omega(t - \tfrac{1}{2}t_0) + \phi_r]\cos(\tfrac{1}{2}\omega t_0 - \phi_r)$$

which vanishes when

$$f_n t_0 = (n + \tfrac{1}{2}) + (\phi_r/\pi). \tag{7.2.10}$$

Here n is a positive integer and f_n the frequencies $\omega/2\pi$ at which phase cancellation occurs. This shows explicitly that the null condition can be achieved over the central portion of the pulses by matching any one of a number of cycles in each echo. One can of course match the pulse envelopes by fine tuning the delay time but, because of pulse shape distortion, this is not a highly accurate method of obtaining the true round trip propagation time t_0. However, if the phase shift ϕ_r can be calculated over the small range of carrier frequencies spanned by a few null values then the transit time can be obtained with high precision and accuracy from equation (7.2.10). Since n must be an integer, measurement of two or more adjacent null frequencies allows the order of interference to be determined. The boundary conditions for a solid sample with quartz transducers bonded to parallel surfaces are reasonably well defined. ϕ_r is close to $\pi/2$ at the resonance frequency of the transducer and its variation around that frequency can be calculated (one must allow for the fluid film used for bonding). An absolute accuracy of 100 PPM or better can be achieved.

To apply this method directly to a fluid sample, one would have to construct transducers with predictable phase shifts on reflection. Rigidly backed quartz expanders operated at resonance would appear to satisfy this requirement. However, a slightly more sophisticated approach, using the apparatus shown in figure 7.5, turns out to be better [14]. In this method a single edge-supported quartz crystal transducer is employed to generate pulses which travel outwards in both directions towards identical reflectors placed distances L_1 and L_2 from the source. The returning echoes are detected by the same transducer. Again two pulses are applied and the delay time between them is adjusted so that the echo of the first pulse travelling the longer path arrives back at the transducer at the same time as the echo of the second pulse travelling the shorter path. The phase difference $\Delta\phi$ between these two returning echoes is then equal to ωt_0, where t_0 is now defined as the transit time over the distance $2(L_2 - L_1)$, and cancellation can occur when $\Delta\phi = (2n + 1)\pi$; the corresponding null frequencies are determined by

$$f_n t_0 = n + \tfrac{1}{2}. \tag{7.2.11}$$

In order to achieve full cancellation the amplitude of the second pulse must be reduced because the first pulse travelling the longer path will suffer greater attenuation. The great advantage of this method, using two identical reflectors, is that phase shifts and pulse shape distortions on reflection cancel. The null frequencies can be determined to a few parts per million and the accuracy should be determined by knowledge of the path difference and by diffraction effects (which will be considered later). Some care is required in the electronics to ensure that, whatever method is used to attenuate the second pulse, phase errors between the two electronic pulses are kept sufficiently small [15].

7.2.4 Multiple-pulse methods

The methods discussed above involve the generation of only one or two pulses in a single measurement. All echoes of the original pulse or pulse pair are allowed to die out before a new measurement is initiated. Pulse superposition methods can also be used with pulses generated in a continuous sequence at a variable repetition rate. Basically there are two methods. In the first, the pulse repetition rate is adjusted manually until echoes of different order all add in phase at the detector [2]. When using a single transducer, a clock can be used to trigger a pulse on every even-numbered cycle and to open a gate to the receiver during every odd-numbered cycle. When the clock period is adjusted to the round trip delay time, all the echoes that have made an odd number of round trips in the sample add in phase at the transducer. As before, cycle-for-cycle matching between pulses can be performed and a precise measurement of the time of flight obtained. Some knowledge of phase shifts on reflection is required for absolute measurements. In the second technique, known as the sing-around method, two transducers are used and each pulse arriving at the receiver is used to trigger a new pulse from the transmitter [16–18]. The time of flight is then obtained from the inverse of the pulse repetition frequency. Since a precision of 1 PPM is not hard to achieve, the method is well suited to the determination of changes in the sound speed (e.g. those accompanying changes in temperature or pressure). For absolute measurements, the delay time in the electronics must be known.

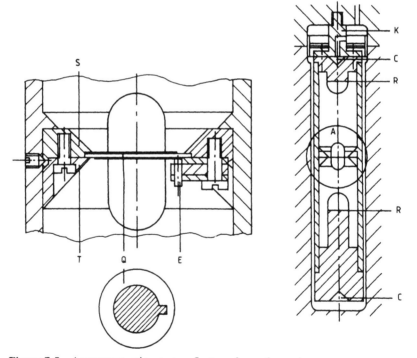

Figure 7.5 Apparatus using two reflectors for pulse–echo measurements on fluid samples at high pressures [14]: K, suspension rod; C, conical cavity; R, reflector; S, support; T, conical ring; Q, transducer; E, silver pin.

7.2.5 Diffraction corrections for pulse measurements

The sound speed (L/t_0) determined from the observed time of flight t_0 of a pulse over the total distance L will be in error by a small amount due to diffraction effects. This error arises from the phase advance of the sound wave relative to a plane wave traversing the same distance. Similarly, the measured attenuation of the pulse will be in error because of beam spreading.

To correct the measured time of flight for diffraction we must add an amount δt given by

$$\delta t = \phi(L)/\omega \qquad (7.2.12)$$

where $\phi(L)$ is the accumulated phase advance over the distance L measured along the axis of the source; this leads to a fractional decrease in u of approximately $\phi u/\omega L$. Generally the receiving transducer is not a point detector, rather it is a circular device equal in diameter to the source (in single-transducer methods, it is the source). Consequently we must calculate the mean acoustic pressure $\langle p_a \rangle$ over the surface of the detector from which the amplitude and phase of the received signal can be obtained [19]. Starting with equation (5.2.15) one can obtain the following expression for the mean acoustic pressure [20]:

$$\langle p_a \rangle = \rho u v_0 \exp(-ikL) \times A \exp(i\phi) \qquad (7.2.13)$$

where

$$A \exp(i\phi) = 1 - \frac{4}{\pi} \int_0^{\pi/2} \exp\{ikz[1 - (1 + 4(b/z)^2 \cos^2\theta)^{1/2}]\}\sin^2\theta \, d\theta$$

$$(7.2.14)$$

and b is the radius of the detector (assumed equal to the radius of the source). The factor $A \exp(i\phi)$ provides a complete description of the free-field diffraction effects for a piston-like source and any parallel detector of the same radius that responds to the mean acoustic pressure over its surface. It remains valid for single-transducer multiple-reflection devices provided that the diameter of the reflection is equal to or greater than that of the transducer. For effective pathlengths greater than a few radii, $\langle p_a \rangle$ becomes a function of the single reduced variable $S = L\lambda/b^2$ and $A \exp(i\phi)$ reduces to

$$A \exp(i\phi) = 1 - \frac{4}{\pi} \int_0^{\pi/2} \exp\left[-\left(\frac{4\pi i}{S}\right)\cos^2\theta\right]\sin^2\theta \, d\theta. \qquad (7.2.15)$$

The diffraction phase advance ϕ is plotted as a function of S in figure 7.6(a) while, in figure 7.6(b), the diffraction losses are shown as $-20 \log_{10}(A)$. These quantities were obtained from equation (7.2.15) by numerical integration; an alternative analytic treatment is given in [21]. In a well designed experiment, the theoretical diffraction correction to the sound speed can often be kept well below 0.01 per cent. This is certainly a wise precaution because perfect piston-like operation of the transducer is unlikely to be achieved in practice and the theory may not always provide a very accurate account of the true diffraction effects. The effect of diffraction (essentially, beam spreading) on the measured absorption can be quite large. For example, the intrinsic attenuation of 1 MHz sound across a 10 cm path in argon gas (at 300 K, 0.1 MPa) is, on a decibel scale, about 17 dB, while with a transducer diameter of 25 mm ($S = 0.2$) there is an additional diffraction loss of about 0.8 dB.

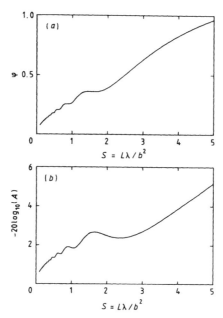

Figure 7.6 Diffraction effects: (*a*) diffraction phase advance ϕ as a function of reduced distance from source $S = L\lambda/b^2$; (*b*) diffraction loss factor A on a decibel scale. For large separations, ϕ approaches $\pi/2$ monotonically while A increases logarithmically.

7.3   Reverberation Methods

Pulse methods work well for absorption measurements in fluids at relatively high frequencies (say, 1 MHz and above) where diffraction losses are relatively small compared with the intrinsic absorption of the fluid. At lower frequencies, measurements of attenuation by pulse methods are subject to large losses from beam spreading and become inaccurate. In that regime, the reverberation technique offers an alternative.

7.3.1 Basic working equations

Essentially, the reverberation method is the transient counterpart of measuring the half-width of a resonance under steady-state conditions. The fluid under study is contained within a suitable resonator and, ideally, a single mode is excited by a transducer. After a steady state is achieved, the source is shut off and the decay of the resonance recorded as a function of time. In the approximation that the walls are rigid and that just a single mode of vibration is involved, the acoustic pressure observed at the detector will decay in proportion to $\exp(-2\pi g t)$, where g is the frequency-domain half-width of the resonance. Thus for times t measured from when the source was shut off, the signal at the detector may be written in terms of real quantities as

$$p_a(t) = p_a(0)\exp(-Dft)\sin(2\pi ft + \phi) \qquad (7.3.1)$$

where $D = 2\pi g/f$ is a logarithmic decrement, f is the natural frequency of the mode and ϕ is a phase angle. We can also define a reverberation time τ by $\tau = 1/Df.$[2] A typical strategy for data collection might involve digitizing the signal at a reading rate in excess of $2f$ over a total period of perhaps five to seven multiples of the reverberation time (about 40 to 60 dB of attenuation). The decrement, and the other parameters in equation (7.3.1) ($p_a(0)$, f and ϕ), can then be fitted to the data. The results of many runs can be averaged to improve the signal-to-noise ratio. This strategy will become impractical at high frequencies because of limitations on the sampling rates that can be achieved. It is then better to demodulate the signal first using a detector tuned to the frequency of the resonance under study and operating with sufficient bandwidth to capture the decay without distortion. A superheterodyne receiver is ideal for this purpose. The output will then be a signal linear in either $\exp(-Dft)$ or $-Dft$ depending upon whether the detector has a linear or a logarithmic response. This can be digitized at a much slower sampling rate (determined by τ^{-1} rather than f) and, since it turns out to be sufficient to determine just the product $Df = \tau^{-1}$, a two-parameter fit will then suffice.

7.3.2 Bulk and surface losses

Since the contribution of bulk absorption to the half-width g is always just $(\alpha/2\pi)u$, irrespective of mode symmetry, the decrement for a resonator with rigid walls (which we denote by D_0) is given by

$$D_0 = (u/f)(\alpha + \beta) \qquad (7.3.2)$$

where β is an equivalent absorption coefficient arising from damping in the boundary layers. Usually, interest has focused on measurement of α

although the experiment could just as well be optimized for measurement of β. Clearly, a sensitive measurement of α requires that the surface losses be made small compared with the volume losses and for this reason the radial modes of spherical resonators have found favour.

For the radial modes of a rigid-walled sphere we have, using equation (3.7.17),

$$\beta = (\pi f/u)(\gamma - 1)(\delta_h/a). \qquad (7.3.3)$$

In the regime where α/f^2 is constant, the ratio α/β is proportional to $f^{3/2}/a$ and surface losses can be made small relative to the bulk losses by resorting to high frequencies and/or large resonators. Unfortunately, pursuing this aim too far involves using modes of high order and consequently invites difficulties with overlap. A second limitation on the resonator's size may arise from the cost of the material required to fill it. In some cases quite small resonators have been used at higher frequencies where all the modes overlap into a continuum [22]. In this case, the resonator can be excited by either a band-limited noise signal or using a frequency-modulated oscillator so that a spectrum of modes in a frequency band $f_0 \pm \Delta f$ is used. One then measures the received signal intensity $\mathfrak{I}$, integrated over the bandwidth, as a function of time. In the continuum approximation, $\mathfrak{I}(t)$ follows a simple decay curve characterized by the value of the decrement at the centre frequency. At sufficiently high frequencies, surface losses are relatively unimportant and the measured decrement is dominated by bulk absorption.

As an example of the use of a single mode, consider the $(0, 10)$ resonance of a 1 l rigid-walled spherical resonator containing water at 300 K and 0.1 MPa; the theoretical resonance frequency is then 115 kHz at which $\alpha = 2.8 \times 10^{-4} \, \mathrm{m}^{-1}$, $\beta = 0.16 \times 10^{-4} \, \mathrm{m}^{-1}$, $D = 2.96 \times 10^{-4}$ and $\tau = 29$ ms. Relatively successful measurements in liquids with absorption coefficients comparable with that of water $(\alpha/f^2 \approx 21 \times 10^{-15} \, \mathrm{s}^2 \mathrm{m}^{-1})$ have been obtained by the use of spherical resonators with volumes in the range from 10 to 100 l or more [23–25]. In one case, measurements were performed using several resonators with different volumes and surface losses accounted for by extrapolating the apparent value of α/f^2 to infinite volume as a linear function of a^{-1} [25]. It is worth noting that, because of its low coefficient of thermal expansion near 300 K, water has an unusually small value of $\gamma - 1$ and that many other fluids will suffer significantly greater damping in the thermal boundary layer; on the other hand, most organic liquids are characterized by values of α/f^2 which are also considerably higher than that of water, so that the relative contributions of bulk and surface damping given above are a useful indication of the general situation for liquids.

If we are interested in the absorption coefficient of a gas then coupling to the shell can be made small and a sensitive measurement is possible.

Consider, for example, a measurement of the absorption coefficient in methane at 300 K and 0.1 MPa, which is dominated by the large bulk viscosity (5.6 mPa s) associated with vibrational relaxation. For the (0, 5) mode in a 1 l resonator, the resonance frequency is then 16.3 kHz but $\alpha = 0.50$ m^{-1}, $\beta = 0.012$ m^{-1}, $D = 0.014$ and $\tau = 4.3$ ms. Fast analogue-to-digital converters currently available could be used to digitize such a decay curve with relative ease. To achieve a good signal-to-noise ratio, one would perform the ring down experiment many times over and calculate the average of the decrements obtained in each run.

7.3.3 Shell motion

For measurements on liquids, a serious practical problem arises from the coupling of fluid and wall motion. The shell stores some of the energy of the system and can have a marked influence on the decrement, shifting it in either direction depending upon the damping of the shell. The situation now is one of coupled oscillators and, assuming that the energies E_{fl} and E_{sh} stored by the fluid and shell stand in a constant ratio during the decay process, the combined decrement is

$$D = D_0 + (D_{sh} - D_0)[E_{sh}/(E_{fl} + E_{sh})] \qquad (7.3.4)$$

where D_{sh} is a decrement associated with the shell [22]. The fraction of the total energy stored by the shell should be linear in the characteristic impedance ρu of the fluid at constant frequency. Usually it appears that $D_{sh} > D_0$ so that the shell increases the apparent damping. Losses from the outer surface can be reduced by evacuating the surroundings of the resonator and by attempting to minimize damping in the supports. Nevertheless, it has often proved impossible to reduce residual losses to a level where absolute measurements of 'small' absorption coefficients can be obtained with any accuracy. However, the method can be applied as a relative one if D_{sh} is taken as a parameter and determined by calibration using a fluid of known absorption [26, 27]. For example, relaxation processes in dilute aqueous solutions have been studied by the reverberation method at frequencies down to around 20 kHz using pure water as the calibration fluid. Since neither the resonance frequencies nor the characteristic impedance are very strong functions of concentration, this would seem to be a reasonable approach. When the resonator is operated in the continuum region, the calibration has also been relied upon to account for the effects of surface (mainly viscous) damping.

The discussion so far has started from the 'ideal' limit of rigid walls but, as indicated in Chapter 3, the opposite limit of perfectly flexible walls has some attraction. There are two reasons for this. First, when one comes to study liquids, it is not possible to obtain a very small specific acoustic wall admittance anyway because the value of the product ρu for the fluid is no longer very much less than that for the wall material. Second, in the limit of pressure release boundary conditions, where the specific acoustic *impedance* z_{sh} of the shell vanishes, there are no thermal or viscous boundary layer losses. The possibility of exploiting this fact does not seem to have been investigated experimentally. However, because the shell must be strong enough to support the weight of the fluid sample it seems unlikely that z_{sh} could be made very small. Let us consider the radial modes of a thin-walled spherical resonator as an example. For these modes the acoustic pressure $p_a(r)$ is proportional to $j_0(K_{0n}r)$, where K_{0n} is chosen such that the boundary condition on the radial fluid speed is satisfied. This condition requires that

$$j_0(K_{0n}a)/j_{0n}'(K_{0n}a) = iz_s \qquad (7.3.5)$$

where z_s is the total specific acoustic impedance of the surface given by

$$z_s = z_{sh}/(1 + z_{sh}y_h) \approx z_{sh}(1 - z_{sh}y_h). \qquad (7.3.6)$$

Since $j_0(x) = \sin(x)/x$, the solutions of equation (7.3.5) may be written for small values of z_{sh} as

$$K_{0n}a = n\pi + iz_{sh} + O(z_{sh}^2) \qquad (7.3.7)$$

where $n = 1, 2, 3 \cdots$ (there is no zero-frequency mode in the limit $z_{sh} \to 0$). Thus, the resonance frequencies $\mathrm{Re}(uK_{0n}/2\pi)$ and, more importantly, the half-widths $\mathrm{Im}(uK_{0n}/2\pi)$ are unaffected by thermal damping at the wall correct to first order in z_{sh}. Similar arguments apply for the non-radial modes which become insensitive to both thermal and viscous effects at the walls. But the questions are: can we make iz_{sh} small enough for these observations to be useful and, if we can, will the frictional losses associated with greatly increased wall motion actually be worse than the losses we were trying to get rid of? An estimation of the order of magnitude of iz_{sh} can be obtained from equation (3.7.18) which, for $2\pi f \approx n\pi(u/a)$, gives

$$iz_{sh} \approx [1 - (f/f_0^2)]/n\pi\rho u^2 C. \qquad (7.3.8)$$

In a 11 Pyrex glass sphere with 1 mm wall thickness (for which $C \approx 4.4 \times 10^{-10}\,\mathrm{Pa}^{-1}$ and $f_0 \approx 20\,\mathrm{kHz}$) filled with water at 300 K and 0.1 MPa, $iz_{sh} \approx 0.18$ at the frequency (13 kHz) of the lowest radial mode. Although a much thinner wall would present considerable difficulties, it is clear that quite small values of the specific acoustic admittance of the wall can be achieved leading to a drastic reduction in the boundary layer losses. Similar results apply for thin-walled metal spheres (which may have lower internal damping then Pyrex) and radiation losses can be excluded as already described. However, the problem of holding the resonator in a loss-free manner would, in the absence of nodal points on its surface, be very difficult to overcome experimentally.

7.3.4 Excitation and detection

Since, with liquid samples, shell motion is quite strongly coupled to the fluid, one may as well take advantage of that fact and attach the transducers to the outer surface of the wall; this avoids having to immerse them in the liquid. Small piezoelectric ceramic elements, cemented to the outer surface and operated at frequencies well below resonance, have been used for this purpose [24–27]. Piezoelectric polymer film should make an excellent alternative. Often a pair of transducers have been employed although a single device, switched between generator and receiver circuits, will suffice.

The disadvantage of a surface-mounted source is that it excites all modes of the resonator with roughly comparable amplitude. Much more efficient ways of exciting pure radial modes in a spherical resonator include placing a small source at the centre of the sphere [28] or guiding sound to the centre along a small tube [29]. These methods exploit the fact that only radial modes have non-zero wavefunctions at the centre of a spherical resonator. Although the source is then placed at the point of highest energy density, where it has the greatest potential to perturb the performance of the resonator, these methods would allow the benefits of radial modes to be extended to higher frequencies.

References

[1] Williams J and Lamb J 1958 *J. Acoust. Soc. Am.* **30** 308
[2] McSkimin H J 1961 *J. Acoust. Soc. Am.* **33** 12
[3] Papadakis E P 1967 *J. Acoust. Soc. Am.* **42** 1045
[4] Tempest W and Parbrook H D 1957 *Acustica* **7** 354
[5] Holmes R, Parbrook H D and Tempest W 1960 *Acustica* **10** 155
[6] Morse P M and Ingard K U 1968 *Theoretical Acoustics* (New York: McGraw-Hill) pp 477–9
[7] Kaye G W C and Laby T H 1973 *Tables of Physical and Chemical Constants* 14th edn (London: Longman) pp 68–9
[8] Kinsler L E, Frey A R, Coppens A B and Sanders J V 1982 *Fundamentals of Acoustics* 3rd edn (New York: Wiley) p 461
[9] Williams R C and Chase C E 1967 *Phys. Rev.* **158** 200
[10] Abrahams B M, Eckstein Y, Ketterson J B and Roach P 1970 *Phys. Rev.* A **1** 250
[11] Kessler L W, Hawley S A and Dunn F 1971 *Acustica* **24** 105
[12] Liebenberg D H, Mills R L and Bronson J C 1974 *J. Appl. Phys.* **45** 741
[13] Petitet J P, Tufeu R and Le Neindre B 1983 *Int. J. Thermophys.* **4** 35
[14] Kortbeek P J, Muringer M J P, Trappeniers N J and Biswas S N 1985 *Rev. Sci. Instrum.* **56** 1269
[15] Kortbeek P J and Schouten J A 1990 *Int. J. Thermophys.* **11** 455

[16] Holbrook R D 1948 *J Acoust. Soc. Am.* **20** 590
[17] Cedrone N P and Curran D R 1951 *J. Acoust. Soc. Am.* **23** 627
[18] Forgacs R L 1960 *J. Acoust. Soc. Am.* **32** 1697
[19] Papadakis E P 1966 *J. Acoust. Soc. Am.* **40** 863
[20] Williams A O Jr 1951 *J. Acoust. Soc. Am.* **23** 1
[21] Bass R 1958 *J. Acoust. Soc. Am.* **30** 602
[22] Lawley L E and Reed R D C 1955 *Acustica* **5** 316
[23] Leonard R W 1948 *J. Acoust. Soc. Am.* **20** 224
[24] Wilson D A and Liebermann L N 1947 *J. Acoust. Soc. Am.* **19** 286
[25] Moen C J 1951 *J. Acoust. Soc. Am.* **23** 62
[26] Kurtze G and Tamm K 1953 *Acustica* **3** 33
[27] Ohsawa T and Wada Y 1967 *Jpn. J. Appl. Phys.* **6** 1351
[28] Keolian R, Garratt S, Maynard J and Rudnick I 1978 *J. Acoust. Soc. Am.*
     64 (S1) 561; 1979 *Bull. Am. Phys. Soc.* **24** 623
[29] Mehl J B and Moldover M R 1981 *J. Chem. Phys.* **74** 4062
[30] Gooberman G L 1968 *Ultrasonics, Theory and Application* (London: English
     University Press) pp 24–5
[31] Kiyohara O, Grolien J-P E and Benson G C 1974 *Can. J. Chem.* **52** 2287

Notes

1. The transform and its inverse are linear operations in the sense that if
   $P(t) = P_1(t) + iP_2(t)$, where P_1 and P_2 are both real-valued functions, then
   $S(\omega) = S_1(\omega) + iS_2(\omega)$ where S_1 is the transform of P_1 and S_2 is the transform
   of P_2. In the present case, $P_1 = w(t)\cos(\omega_0 t)$ and $P_2 = w(t)\sin(\omega_0 t)$. If $p(t)$ is
   a real-valued function then its transform $s(\omega)$ must have the property that
   $s(-\omega) = s^*(\omega)$; conversely, a function $s(\omega)$ with that property must have as its
   inverse transform a real-valued function. Consequently, since $\mathfrak{I}(-\omega) = \mathfrak{I}^*(\omega)$
   and $S_1(\omega)$ and $S_2(\omega)$ both have real-valued inverse transforms, the inverse
   transforms of $\mathfrak{I}(\omega)S_1(\omega)$ and $\mathfrak{I}(\omega)S_2(\omega)$ are also both real-valued functions.
2. Other definitions of reverberation time are used in connection with room
   acoustics.

Appendix 1

Properties of Some Mathematical Functions

The purpose of this appendix is to give some properties, especially those required to evaluate expressions in the text, of cylindrical and spherical Bessel functions and of the associated Legendre functions and the spherical harmonics formed from them.

A1.1   Cylindrical Bessel Functions $J_m(z)$

The general solution of the second-order differential equation

$$(d^2 F/dz^2) + (1/z)(dF/dz) + [1 - (m/z)^2] F = 0 \qquad (A1.1)$$

may be written $F(z) = A J_m(z) + B N_m(z)$, where A and B are arbitrary constants, $N_m(z)$ is the cylindrical Neumann function of order m, and $J_m(z)$ is the cylindrical Bessel function of order m. When m is zero or a positive integer, $J_m(z)$ may be obtained from the infinite series

$$J_m(z) = \sum_{s=0}^{\infty} \frac{(-1)^s (z/2)^{m+2s}}{s!(m+s)!} \qquad (A1.2)$$

or by the integral representation

$$J_m(z) = \frac{1}{2\pi i^m} \int_0^{2\pi} \exp[iz \cos(w)] \cos(mw) \, dw \qquad (A1.3)$$

while for large values of z, the following asymptotic form applies:

$$J_m(z) \xrightarrow[z \to \infty]{} (2/\pi z)^{1/2} \cos[z - (2m+1)\pi/4] . \qquad (A1.4)$$

The functions of negative order are given by

$$J_{-m}(z) = (-1)^m J_m(z). \tag{A1.5}$$

The following recursion formulae are also useful in evaluating $J_m(z)$ and its first derivative:

$$J_{m-1}(z) + J_{m+1}(z) = (2m/z) J_m(z) \tag{A1.6}$$

$$(d/dz) J_m(z) = \tfrac{1}{2} [J_{m-1}(z) - J_{m+1}(z)]. \tag{A1.7}$$

Finally some useful integrals of Bessel functions are

$$\int J_0(z) z \, dz = z J_1(z) \tag{A1.8}$$

$$\int J_1(z) \, dz = -J_0(z) \tag{A1.9}$$

$$\int J_0^2(z) z \, dz = (z^2/2)[J_0^2(z) + J_1^2(z)] \tag{A1.10}$$

$$\int J_m^2(z) z \, dz = (z^2/2)[J_m^2(z) - J_{m-1}(z) J_{m+1}(z)] \tag{A1.11}$$

$$\int J_m(az) J_m(bz) z \, dz = [z/(a^2 - b^2)]$$
$$\times [b J_m(az) J_{m-1}(bz) - a J_m(bz) J_{m-1}(az)]$$
$$(a \neq b). \tag{A1.12}$$

The turning points $z = \chi_{mn}$ of the mth-order function are the solutions of $(d/dz) J_m(z) = 0$ and may be obtained numerically with the aid of equations (A1.2) and (A1.7). In Chapter 3 we assumed that the set of functions $J_m(\chi_{mn} z)$, $n = 1, 2, 3 \cdots$, obeys the orthogonality condition

$$\int_0^1 J_m(\chi_{mn} z) J_m(\chi_{mn'} z) z \, dz = \tfrac{1}{2} [1 - (m/\chi_{mn})^2] J_m^2(\chi_{mn}) \delta_{nn'}. \tag{A1.13}$$

It is now a simple matter to demonstrate this: from equations (A1.6) and (A1.7) we have

$$J_{m+1}(\chi_{mn}) = J_{m-1}(\chi_{mn}) \tag{A1.14}$$

$$J_{m-1}(\chi_{mn}) = (m/\chi_{mn}) J_{mn}(\chi_{mn}) \tag{A1.15}$$

which when substituted in (A1.11) and (A1.12) yield the required results. Equation (A1.13) may be combined with the properties of the trigonometric functions to obtain the normalization constant Λ_N^0 of equation (3.6.9).

A1.2 Spherical Bessel Functions $j_l(z)$

The differential equation

$$(d^2 F/dz^2) + (2/z)(dF/dz) + [1 - l(l+1)/z^2] F = 0 \qquad (A2.1)$$

has a general solution that may be written $F(z) = Aj_l(z) + Bn_l(z)$, where A and B are arbitrary constants, $n_l(z)$ is the spherical Neumann function of order l and $j_l(z)$ is the spherical Bessel function of order l. The particular solution $h_l(z) = j_l(z) + in_l(z)$, which is called the spherical Hankel function of order l, is given by the finite series

$$h_l(z) = (i^{-l}/iz)\exp(iz) \sum_{s=0}^{l} \frac{(l+s)!}{s!(l-s)!} \left(\frac{i}{2z}\right)^s. \qquad (A2.2)$$

Thus, $j_l(z)$ and $n_l(z)$ can be obtained numerically as the real and imaginary parts of equation (A2.2). Explicitly, for the first few values of l, we have

$$
\begin{aligned}
j_0(z) &= z^{-1} \sin(z) \\
j_1(z) &= z^{-2} \sin(z) - z^{-1} \cos(z) \\
j_2(z) &= (3z^{-3} - z^{-1})\sin(z) - 3z^{-2} \cos(z) \\
n_0(z) &= -z^{-1} \cos(z) \\
n_1(z) &= -z^{-1} \sin(z) - z^{-2} \cos(z) \\
n_2(z) &= -(3z^{-2} - z^{-1})\cos(z) - 3z^{-3} \sin(z).
\end{aligned}
\qquad (A2.3)
$$

In addition, for large values of z the following asymptotic form of $j_l(z)$ applies:

$$j_l(z) \xrightarrow[z \to \infty]{} (1/z)\cos[z - \tfrac{1}{2}(l+1)\pi]. \qquad (A2.4)$$

The functions of negative order may be obtained from the relation

$$h_{-l}(z) = i(-1)^{l-1} h_{l-1}(z). \qquad (A2.5)$$

Some important properties of the spherical Bessel functions of integer order are the recursion relations

$$j_{l-1}(z) + j_{l+1}(z) = [(2l+1)/z] j_l(z) \qquad (A2.6)$$

$$(d/dz)j_l(z) = (2l+1)^{-1} [lj_{l-1}(z) - (l+1)j_{l+1}(z)] \qquad (A2.7)$$

and the integrals

$$\int j_0(z)z^2 \, dz = z^2 j_1(z) \qquad (A2.8)$$

$$\int j_1(z) \, dz = -j_0(z) \qquad (A2.9)$$

$$\int j_0^2(z)z^2 \, \mathrm{d}z = (z^3/2)[j_0^2(z) + n_0(z)j_1(z)] \qquad (A2.10)$$

$$\int j_l^2(z)z^2 \, \mathrm{d}z = (z^3/2)[j_l^2(z) - j_{l-1}(z)j_{l+1}(z)] \qquad (A2.11)$$

$$\int j_l(az)j_l(bz)z^2 \, \mathrm{d}z = [z^2/(a^2 - b^2)]$$

$$\times [bj_l(az)j_{l-1}(bz) - aj_l(bz)j_{l-1}(az)]$$

$$(a \neq b). \quad (A2.12)$$

These properties can be used to prove the orthogonality relation

$$\int_0^1 j_l(\nu_{ln}z)j_l(\nu_{ln'}z)z \, \mathrm{d}z = \tfrac{1}{2}[1 - l(l+1)/\nu_{ln}^2]j_l^2(\nu_{ln})\delta_{nn'} \qquad (A2.13)$$

for the set of functions $j_l(\nu_{ln}z)$, $n = 1, 2, 3 \cdots$, assumed in Chapter 3. Here, ν_{ln} are the solutions of $(\mathrm{d}/\mathrm{d}z)j_l(z) = 0$. Equation (A2.13) follows from (A2.11) and (A2.12) because, using (A2.6) and (A2.7), we have

$$j_{l-1}(\nu_{ln}) = [(l+1)/\nu_{ln}]j_l(\nu_{ln}) \qquad (A2.14)$$

$$j_{l+1}(\nu_{ln}) = (l/\nu_{ln})j_l(\nu_{ln}). \qquad (A2.15)$$

A1.3   The Associated Legendre Functions $P_l^{|m|}(\eta)$

The solution of the differential equation

$$(\mathrm{d}^2 F/\mathrm{d}\theta^2) + (\cos\theta/\sin\theta)(\mathrm{d}F/\mathrm{d}\theta) + [l(l+1) - (m/\sin\theta)^2]F = 0 \quad (A3.1)$$

that is finite in the interval $0 \leqslant \theta \leqslant \pi$ may be written $F(\theta) = AP_l^{|m|}(\cos\theta)$, where A is an arbitrary constant and $P_l^{|m|}(\cos\theta)$ is the associated Legendre function of order l. Here l is a positive integer or zero and m is restricted to values $0, \pm 1, \pm 2, \cdots, \pm l$. The only property of the associated Legendre functions of which we make use is the relation

$$\int_{-1}^1 P_l^{|m|}(\eta)P_{l'}^{|m|}(\eta) \, \mathrm{d}\eta = 2(2l+1)^{-1}\left(\frac{(l+|m|)!}{(l-|m|)!}\right)\delta_{ll'} \qquad (A3.2)$$

where $\eta = \cos\theta$. The spherical harmonic $Y_{lm}(\theta, \xi)$ used in the text is defined by

$$Y_{lm}(\theta, \xi) = [\cos(m\xi) + \sin(m\xi)]P_l^{|m|}(\cos\theta) \qquad (A3.3)$$

and since

$$\int_0^{2\pi} [\cos(m\xi) + \sin(m\xi)][\cos(m'\xi) + \sin(m'\xi)] \, \mathrm{d}\xi = 2\pi\delta_{mm'} \quad (A3.4)$$

Y_{lm} obeys the orthogonality condition

$$\int_0^{2\pi} \int_0^{\pi} [Y_{lm}(\theta, \xi)]^2 \sin \theta \; \mathrm{d}\theta \; \mathrm{d}\xi = 4\pi(2l + 1)^{-1} \left(\frac{(l + |m|)!}{(l - |m|)!} \right) \delta_{mm'} \delta_{ll'}. \tag{A3.5}$$

This may be combined with equation (A2.13) to obtain the normalization constant Λ_N^0 of equation (3.7.12).

General References

Abramowitz M and Stegun I A (eds) 1964 *Handbook of Mathematical Functions* (Washington, DC: US Government Printing Office) Ch. 8–10

Morse P M and Feshbach H 1953 *Methods of Theoretical Physics* (New York: McGraw-Hill) part II, pp 1321–6, 1573–6

Morse P M and Ingard K U 1968 *Theoretical Acoustics* (New York: McGraw-Hill) pp 210, 332–8

Olver F W J (ed.) 1960 *Royal Society Mathematical Tables Volume 7: Bessel Functions part III, zeros and associated values* (Cambridge: Cambridge University Press)

Appendix 2

Notation

It has not been possible to employ a unique symbol for every quantity used; however, exact duplication has been kept to a minimum through decoration of characters. The following list gives the main symbols used in the text.

Roman, Italic and Script Characters

a	Radius of sphere, circular membrane or plate; specific affinity
$A, B, C \cdots$	Chemical species
A, B, C	Amplitudes of wavefunctions
A_i	$(i = 0, 1, 2 \cdots)$ coefficients of p^i in expansion of u^2
A_m	Molar Helmholtz energy
A_{NM}	Surface integrals in boundary shape perturbation theory
$\mathcal{A}$	Area
b	Radius of cylinder
B, C, D	Second, third, fourth virial coefficient
B	Magnitude of magnetic flux density
B	Matrix of normalized surface integrals
c	Molecular speed
c	Matrix of elastic moduli
c_p, c_V	Isobaric, isochoric specific heat capacities
C_0, C_1, C_2	Electric capacitances (also $C_\mathrm{A}, C_\mathrm{B}, C_\mathrm{C}$)
C_int	Contribution of internal modes to heat capacity
C_n	Contribution of nth internal mode to heat capacity
C_p, C_V	Isobaric, isochoric heat capacities
C_rot	Rotational contribution to heat capacity
C_vib	Vibrational contribution to heat capacity
d	Thickness of plate, membrane or gap
d	Matrix of piezoelectric strain coefficients
D	Binary diffusion coefficient; logarithmic decrement

240

D	Electric flux density
D_h	Thermal diffusivity $\varkappa/\rho c_p$
D_s	Viscous diffusivity η/ρ
D_v	Modified viscous diffusivity $(4\eta/3\rho) + (\eta_b/\rho)$
e	Base of natural logarithms $(2.718\,28\cdots)$
e_r, e_θ, e_z	Unit base vectors in cylindrical coordinate system
e_r, e_θ, e_ξ	Unit base vectors in spherical coordinate system
e	Matrix of piezoelectric stress coefficients
E	Thermodynamic energy
E_{int}	Contribution of internal modes to thermodynamic energy
E_{tr}	Contribution of translational modes to thermodynamic energy
E	Electric field
f	Frequency
f_n	Resonance frequency of membrane or plate
f_N	Resonance frequency of Nth normal mode
$F; F$	Force; magnitude of force
F_N	Complex resonance frequency of Nth normal mode
g	Specific Gibbs energy
g_N	Resonance half-width of Nth normal mode
g_b	Half-width contribution from bulk absorption
g_h, g_s	Half-width contributions from thermal, shear boundary layers
g_{sh}	Half-width contribution from shell motion
$g(r\,|\,r_0)$	Green function for free space
G	Gain of amplifier
$G(r\,|\,r_0)$	General Green function
$G_\omega(r\,|\,r_0)$	Green function for frequency $\omega/2\pi$
$G_N(r\,|\,r_0)$	Green function for $k = K_N$
$G_N'(r\,|\,r_0)$	Green function for $k = K_N$ neglecting the Nth term in its normal-mode expansion
G_m	Molar Gibbs energy
h	Specific enthalpy
$h_l(z)$	Spherical Hankel function of order l
h	Matrix of piezoelectric coefficients; unit vector in direction of propagation of a plane wave
i	$\sqrt{-1}$
i, j, k	x, y, z-pointing unit base vectors
I	Electric current
$\mathscr{I}$	Acoustic intensity
$I_m(z)$	Hyperbolic cylindrical Bessel function of order m
$\mathscr{I}$	Nine-component tensor with elements $\mathscr{I}_{ij} = \delta_{ij}$
$j_l(z)$	Spherical Bessel function of order l
J_h	Heat current density
J_P	Momentum current density
$J_m(z)$	Cylindrical Bessel function of order m

$\mathscr{J}$	Transverse contribution to viscous stress tensor
k	Propagation constant; Boltzmann's constant
k_{KH}	Kirchhoff–Helmholtz tube propagation constant
k_M	ω/u_M
k_P	ω/u_P
$-k_N^2$	Eigenvalue for idealized cavity
$-K_N^2$	Perturbed eigenvalue
K_t	Electromechanical coupling constant
K_T	Thermal diffusion ratio
l	Normal mode index; length; half thickness of piezoelectric plate
L	Avogadro's constant; length of cylinder or tube
L'	Effective length of open-ended tube
L_1, L_2, L_3	Electric inductances (also L_A, L_B, L_C)
L_n^r, L_n^a	Resonance, anti-resonance lengths of interferometer
$L_n^{(A)}$	Apparent resonance length determined from admittance circle
m	Mass
M	Molar mass
$\bar{n}$	Number density
n	Unit vector normal to a surface
$n_l(z)$	Spherical Neumann function of order l
N	Trio of indices identifying a normal mode (M also used)
$N_m(z)$	Cylindrical Neumann function of order m
p	Hydrostatic pressure
p_a	Acoustic pressure
p_i	Instantaneous acoustic pressure; amplitude of incident sound pressure wave
P	Power; momentum
$P(t)$	Pulse function
$P_l(z)$	Legendre function of order l
$P_l^{|m|}(z)$	mth associated Legendre function of order l
q	General Cartesian coordinate; heat per unit mass
Q	Quality factor
Q_N	Quality factor of Nth normal mode
r	Radial coordinate; intermolecular separation; characteristic acoustic impedance ρu
r	Position vector
r_S	Position vector for a surface location
r_0	Position vector of elementary source
R	Gas constant; distance from elementary source $|r - r_0|$; ratio (b/L) for cylinder
R_m	Mechanical resistance
R_1, R_2, R_3	Electrical resistances (also R_A, R_B, R_C)
s	Specific entropy; stiffness

s	Elastic compliance matrix
S	Entropy; surface area; diffraction parameter $(L\lambda/b^2)$
S_ω	Source strength for frequency $\omega/2\pi$
$S(\omega)$	Fourier transform of pulse function $P(t)$
$\mathscr{S}$	Strain tensor
t	Time
T	Temperature
T_a	Acoustic temperature
$\mathfrak{J}$	Tension per unit length around perimeter of membrane
$\mathscr{T}$	Stress tensor
u	General sound speed; phase speed
u_0	Zero-frequency sound speed
u_g	Group speed
u_s	Signal speed
u_∞	Hypothetical infinite frequency sound speed
u_M	Speed of transverse waves on membrane
u_P	Speed of transverse waves on plate
$U(r)$	Intermolecular pair potential energy function
v	Speed
v	Velocity; fluid velocity
v_l	Longitudinal (irrotational) component of fluid velocity
v_r	Rotational (transverse) component of fluid velocity
V	Volume; potential difference
w	Energy density; complex resonance signal
$w(t)$	Gating function
x	Cartesian coordinate; mole fraction
y	Cartesian coordinate; specific acoustic admittance $\rho u Y$ (β in §3.5.2)
y_h, y_s	Specific acoustic admittance of thermal, shear boundary layers
y_o	Specific acoustic admittance of opening
y_{sh}	Specific acoustic admittance of shell wall
y_0	Specific acoustic admittance of boundary or of waveguide at $z = 0$
y_L	Specific acoustic admittance of waveguide at $z = L$
Y	Acoustic admittance; Young's modulus
Y_m	Mechanical admittance
$Y_{lm}(\theta, \xi)$	Spherical harmonic
z	Cartesian coordinate; specific acoustic impedance $Z/\rho u$
z_r	Specific acoustic radiation impedance
Z	Compression factor
Z_a	Acoustic impedance
Z_e	Electrical impedance
Z_m	Mechanical impedance
Z_r	Acoustic radiation impedance

Greek Characters

α	Absorption coefficient
$\alpha, \beta, \gamma, \delta$	Critical exponents
α_{cl}	Classical absorption coefficient
α_{rel}	Relaxation absorption coefficient
α_t^2	Factor relating mechanical and electrical admittance
α_{KH}	Kirchhoff–Helmholtz tube attenuation constant
β	Thermal pressure coefficient $(\partial p / \partial T)_\rho$; effective absorption coefficient arising from surface losses
$\boldsymbol{\beta}$	Dielectric impermeability matrix
$\beta_a, \gamma_a, \delta_a$	Second, third, fourth acoustic virial coefficients
γ	Ratio of heat capacities C_p / C_V; dimensionless frequency ω / ω_0
γ_{eff}	Effective value of C_p / C_V
γ_ω	Effective value of C_p / C_V for harmonic cycle at frequency $\omega / 2\pi$
Γ	Reduced propagation constant $k u_0 / \omega$
Γ_∞	Reduced propagation constant $k u_\infty / \omega$
Γ_h, Γ_s	Reduced propagation constant for thermal, shear mode
Γ_t, Γ_r	Factors of Γ pertaining to translational, relaxation effects
δ_h, δ_s	Thermal, shear penetration lengths in fluid
δ_w	Thermal penetration length in wall material
δ_{ij}	Dirac delta function $\delta_{ij} = 0$ when $i \neq j$, $\delta_{ij} = 1$ when $i = j$
Δ	C_{int} / C_p
Δ_n	C_n / C_p
Δ_∞	$C_{int} / (C_p - C_{int})$
Δ_{rot}	C_{rot} / C_p
$\Delta_1, \Delta_2, \Delta_3$	Chemical enhancement factors
$\Delta f_h, \Delta f_s$	Resonance frequency shifts arising from thermal, shear boundary layers
Δf_{sh}	Resonance frequency shift arising from shell motion
$\Delta L_n^r, \Delta L_n^a$	Resonance, anti-resonance half-widths for interferometer
$\Delta L_n^{(A)}$	Apparent resonance half-width determined from admittance circle
ε	Dielectric permittivity; parameter describing geometric imperfection; non-linearity parameter
ε_r	Relative dielectric permittivity (dielectric constant)
ε_0	Dielectric permittivity of free space
$\boldsymbol{\varepsilon}$	Dielectric permittivity matrix
ζ	Collision number τ / τ_c
ζ_{rot}	Rotational collision number
ζ_{vib}	Vibrational collision number
η	Coefficient of shear viscosity
η_b	Coefficient of bulk viscosity
θ	Polar angle; dimensionless time variable $\omega_0 t$

$\varkappa$	Coefficient of thermal conductivity
$\varkappa_w$	Thermal conductivity of wall material
$\varkappa_S$	Isentropic compressibility $\rho^{-1}(\partial\rho/\partial p)_S$
$\varkappa_T$	Isothermal compressibility $\rho^{-1}(\partial\rho/\partial p)_T$
λ	Wavelength
$\lambda_0, \lambda_\infty$	u_0/f, u_∞/f
λ_m	Mean free path
λ_n	nth characteristic value of B matrix
Λ_N	Normalization constant of Nth normal mode
Λ_N^0	Unperturbed normalization constant of Nth normal mode
μ	Absorption per wavelength $\alpha\lambda$
μ_{max}	Maximum absorption per wavelength
ν_n	Constant for nth resonance frequency of edge-clamped circular plate
ν_{ln}	nth turning point of the spherical Bessel function of order l
ξ	Azimuthal angle; surface displacement; extent of reaction; specific acoustic conductance
π	Ratio of circumference to diameter of circle (3.141 59$\cdots$)
ρ	Mass density
ρ_a	Acoustic density fluctuations
ρ_n	Amount-of-substance density
ρ_S	Mass per unit area
σ	Surface charge density; specific acoustic susceptance; Poisson's ratio
τ	Isothermal relaxation time; reverberation time; pulse width
τ_h, τ_s	Characteristic thermal and shear relaxation times, D_h/u_0^2 and D_s/u_0^2
τ'	Adiabatic constant-volume relaxation time
τ''	Adiabatic constant-pressure relaxation time
τ_c	Mean time between collisions
τ_r	Rise time
τ_v	Modified viscous relaxation time (D_v/u_0^2)
τ_{rot}	Rotational relaxation time
τ_{vib}	Vibrational relaxation time
ϕ	Phase angle; $-(\partial a/\partial\xi)_{p,T}$ with a = specific affinity
$\phi(r)$	Wavefunction
$\Phi(r)$	Wavefunction
Φ_0, Φ_L	Tube impedance parameters
χ_{mn}	nth turning point of the cylindrical Bessel function of order m
χ_{Re}, χ_{Tr}	Reflection, transmission coefficient at an interface
χ_r, χ_t	Reflection coefficient at reflector, transducer of interferometer
$\psi(r)$	Wavefunction
ψ_{mn}	nth zero of $J_m(z)$
Ψ	Velocity potential

ω	Angular frequency $2\pi f$
ω_{max}	Angular frequency at which $\mu = \mu_{max}$

Subscripts and Superscripts

a	Acoustic quantity
eq	Equilibrium value
h	Thermal mode
l, r	Longitudinal, rotational component
m	Molar quantity
p	Propagational mode
pg	Perfect gas
s	Shear mode

Index